Principles

of Photochemistry

Principles and Applications of Photochemistry

Brian Wardle
Manchester Metropolitan University, Manchester, UK

A John Wiley & Sons, Ltd., Publication

This edition first published 2009
© 2009 John Wiley & Sons, Ltd

Registered office
John Wiley & Sons Ltd, The Atrium, Southern Gate, Chichester, West Sussex, PO19 8SQ, United
Kingdom

For details of our global editorial offices, for customer services and for information about how to apply
for permission to reuse the copyright material in this book please see our website at www.wiley.com.

Library of Congress Cataloging-in-Publication Data

Wardle, Brian.
  Principles and applications of photochemistry / Brian Wardle.
    p. cm.
  Includes bibliographical references and index.
  ISBN 978-0-470-01493-6 (cloth) – ISBN 978-0-470-01494-3 (pbk. : alk. paper)
1. Photochemistry. I. Title.
  QD708.2.W37 2009
  541′.35–dc22

                                                                                2009025968

A catalogue record for this book is available from the British Library.

ISBN CLOTH 9780470014936, PAPER 9780470014943

Set in 10.5 on 13 pt Sabon by Toppan Best-set Premedia Limited

Dedication

To my family, past and present and to my tutors, all of whom shed the light.

And God said, Let there be light and there was light

And God saw the light, that it was good and God divided the light from the darkness

Genesis 1,3

Contents

Preface

Photochemistry is the branch of chemistry which relates to the interactions between matter and photons of visible or ultraviolet light and the subsequent physical and chemical processes which occur from the electronically excited state formed by photon absorption.

The aim of this book is to provide an introduction to the principles and applications of photochemistry and it is generally based on my lectures to second and third-year undergraduate students at Manchester Metropolitan University (MMU).

Chapters 1 and 2 give a general introduction to the concepts of light and matter and to their interaction resulting in electronically excited states. Chapters 3 to 6 relate to processes involving physical deactivation of the electronically excited states. Chapter 7 provides an overview of the chemical properties of excited states including their reaction pathways and the differences between photochemical reactions and the so-called 'thermal' reactions. Chapters 8 and 9 relate to the photochemical reactions of two of the more interesting groups of organic compounds, namely alkenes and carbonyl compounds. Here the photochemical reactions provide an important extension to the reactions the ground state species. Chapter 10 considers some mechanistic aspects of photochemical reactions and looks at some techniques employed in this field. Chapters 11 and 12 cover areas where outstanding progress has been made in recent years. Chapter 11 considers semiconductor photochemistry whereas in Chapter 12 an introduction to supramolecular photochemistry is presented.

The author gratefully acknowledges the help and advice given by various colleagues and friends, particularly Prof. Norman Allen, Dr Paul Birkett, Dr Michelle Edge, Dr Paul Monk, Dr Christopher Rego, (all MMU) and Dr Michael Mortimer (Open University).

Brian Wardle
Manchester Metropolitan University

1

Introductory Concepts

AIMS AND OBJECTIVES

After you have completed your study of all the components of Chapter 1, you should be able to:

- Understand the concept of the quantised nature of light and matter and be able to draw simple diagrams showing quantised energy levels in atoms and molecules.
- Relate the wavelength of electromagnetic radiation to its frequency and energy.
- Understand the relationship between the wavelength of electromagnetic radiation absorbed by a sample and its potential to produce chemical change.
- Understand how absorption, spontaneous emission and stimulated emission occur in matter–light interactions.
- Explain how quantum mechanics has led to the concept of atomic and molecular orbitals.
- Use and interpret simple atomic and molecular orbital energy diagrams.
- Describe the relative merits as light sources of mercury lamps and lasers.
- Describe the mode of action of a laser and the characteristic properties of laser light.
- Distinguish between electronic configuration and electronic state.
- Recognise experimental situations in which lasers are essential and those in which mercury lamps are more appropriate.

Principles and Applications of Photochemistry   Brian Wardle
© 2009 John Wiley & Sons, Ltd

- Understand the importance of quantum yield as a measure of the efficiency of a photoreaction.

1.1 THE QUANTUM NATURE OF MATTER AND LIGHT

Photochemical reactions occur all around us, being an important feature of many of the chemical processes occurring in living systems and in the environment. The power and versatility of photochemistry is becoming increasingly important in improving the quality of our lives, through health care, energy production and the search for 'green' solutions to some of the problems of the modern world. Many industrial and technological processes rely on applications of photochemistry, and the development of many new devices has been made possible by spin-off from photochemical research. Important and exciting light-induced changes relevant to everyday life are discussed throughout this text.

Photochemistry is the study of the chemical reactions and physical changes that result from interactions between matter and visible or ultraviolet light.

The principal aim of this introduction is to familiarise the reader with basic ideas relating to light and matter and the interaction between them. Quantum mechanics underpins an understanding of the nature of both light and matter, but a rigorous treatment of quantum theory involves complex mathematical analysis. In order to make the ideas of quantum mechanics available to a wider readership, conceptually simple models are presented.

The development of the **quantum theory** in the early twentieth century allowed predictions to be made relating to the properties and behaviour of matter and light. The electrons in matter have both wavelike and particle-like properties, and quantum theory shows that the energy of matter is **quantised**; that is, only certain specific energies are allowed.

The quantised energy levels of matter have a separation that is of the same order as the energy of visible or ultraviolet light. Thus the absorption of visible or ultraviolet light by matter can excite electrons to higher energy levels, producing electronically-excited species.

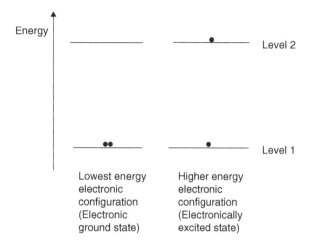

Figure 1.1 Quantised energy levels in matter, where an electron (•) may be found in either of the two energy levels shown

According to the quantum theory, light is also quantised. The absorption or emission of light occurs by the transfer of energy as **photons**. These photons have both wavelike and particle-like properties and each photon has a specific energy, E, given by **Planck's law**:

$$\boxed{E = h\nu}$$

where h is **Planck's constant** (6.63×10^{-34} Js) and ν is the **frequency** of oscillation of the photon in units of s^{-1} or Hertz (Hz).

> The term 'hν' is used in equations for photophysical and photochemical processes to represent a photon.

For example, for a molecule R in its ground state which absorbs a photon to produce an electronically-excited molecule, R*, we may write the process as:

$$R + h\nu \rightarrow R^*$$

Each photon oscillates with wavelength λ, where $\lambda = c/\nu$ and where c is the speed of light. Thus:

$$E = h\nu = hc/\lambda$$

This equation demonstrates the important properties relating to the energy of photons:

> **The energy of a photon is proportional to its frequency and inversely proportional to its wavelength.**

The units most commonly used are:

- J or kJ for the energy of a photon. The energy of one mole of photons (6.02×10^{23} photons) is called an **einstein** and is measured in units of kJ mol^{-1}. One einstein of light of wavelength λ is given by $N_A hc/\lambda$, where N_A is the **Avogadro constant** (6.02×10^{23} mol^{-1}). Sometimes energy is measured in **electronvolts** (eV), where 1 eV = 1.602×10^{-19} J.
- s^{-1} or Hz for frequency, where 1 Hz = 1 s^{-1}.
- nm (nanometre) or Å (angstrom) for wavelength, where 1 nm = 10^{-9} m and 1 Å = 10^{-10} m.

In some literature accounts, the term **wave number** ($\bar{\nu}$) is used. This is the number of wavelengths per centimetre, and consequently wave number has units of reciprocal centimetres (cm^{-1}).

Table 1.1 shows the properties of visible and ultraviolet light.

> **The production of the electronically-excited state by photon absorption is the feature that characterises photochemistry and separates it from other branches of chemistry.**

Table 1.1 Properties of visible and ultraviolet light

Colour	λ/nm	$\nu/10^{14}$ Hz	$\bar{\nu}/10^4$ cm^{-1}	E/kJ mol^{-1}
red	700	4.3	1.4	170
orange	620	4.8	1.6	193
yellow	580	5.2	1.7	206
green	530	5.7	1.9	226
blue	470	6.4	2.1	254
violet	420	7.1	2.4	285
ultraviolet	<300	>10.0	>3.3	>400

Figure 1.2 The process of light absorption

Sometimes electronic excitation can result in chemical changes, such as the fading of dyes, photosynthesis in plants, suntans, or even degradation of molecules. On other occasions, the electronically-excited state may undergo deactivation by a number of physical processes, either resulting in emission of light (luminescence) or conversion of the excess energy into heat, whereby the original ground state is reformed. Electronically-excited states can also interact with ground-state molecules, resulting in energy-transfer or electron-transfer reactions provided certain criteria are met.

There are three basic processes of light–matter interaction that can induce transfer of an electron between two quantised energy states:

1. In **absorption** of light, a photon having energy equal to the energy difference between two electronic states can use its energy to move an electron from the lower energy level to the upper one, producing an electronically-excited state (Figure 1.2). The photon is completely destroyed in the process, its energy becoming part of the total energy of the absorbing species.

   Two fundamental principles relating to light absorption are the basis for understanding photochemical transformations:
   - The **Grotthuss–Draper law** states that only light which is absorbed by a chemical entity can bring about photochemical change.
   - The **Stark–Einstein law** states that the primary act of light absorption by a molecule is a one-quantum process. That is, for each photon absorbed only one molecule is excited. This law is obeyed in the vast majority of cases but exceptions occur when very intense light sources such as lasers are used for irradiation of a sample. In these cases, concurrent or sequential absorption of two or more photons may occur.
2. **Spontaneous emission** occurs when an excited atom or molecule emits a photon of energy equal to the energy difference between the two states without the influence of other atoms or molecules (Figure 1.3(a)):

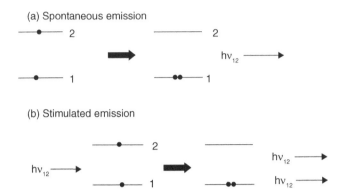

Figure 1.3 The processes of (a) spontaneous emission and (b) stimulated emission

$$R^* \rightarrow R + h\nu_{12}$$

Light is emitted from the bulk material at random times and in all directions, such that the photons emitted are out of phase with each other in both time and space. Light produced by spontaneous emission is therefore called **incoherent light.**

3. **Stimulated emission** occurs when a photon of energy equal to the energy difference between the two states interacts with an excited atom or molecule (Figure 1.3(b)):

$$R^* + h\nu_{12} \rightarrow R + 2h\nu_{12}$$

The photons produced by stimulated emission are in phase with the stimulating photons and travel in the same direction; that is, the light produced by stimulated emission is **coherent light.** Stimulated emission forms the basis of laser action.

1.2   MODELLING ATOMS: ATOMIC ORBITALS

Erwin Schrödinger developed an equation to describe the electron in the hydrogen atom as having both wavelike and particle-like behaviour. Solution of the **Schrödinger wave equation** by application of the so-called **quantum mechanics** or **wave mechanics** shows that electronic energy levels within atoms are quantised; that is, only certain specific electronic energy levels are allowed.

Solving the Schrödinger wave equation yields a series of mathematical functions called **wavefunctions**, represented by Ψ (Greek letter psi), and their corresponding energies.

The square of the wavefunction, Ψ^2, relates to the probability of finding the electron at a particular location in space, with **atomic orbitals** being conveniently pictured as boundary surfaces (regions of space where there is a 90% probability of finding the electron within the enclosed volume).

In this quantum mechanical model of the hydrogen atom, three **quantum numbers** are used to describe an atomic orbital:

- The **principal quantum number**, n, can have integral values of 1, 2, 3, etc. As n increases, the atomic orbital is associated with higher energy.
- The **orbital angular-momentum quantum number**, ℓ, defines the shape of the atomic orbital (for example, s-orbitals have a spherical boundary surface, while p-orbitals are represented by a two-lobed shaped boundary surface). ℓ can have integral values from 0 to (n − 1) for each value of n. The value of ℓ for a particular orbital is designated by the letters s, p, d and f, corresponding to ℓ values of 0, 1, 2 and 3 respectively (Table 1.2).
- The **magnetic quantum number**, m_l, describes the orientation of the atomic orbital in space and has integral values between −l and +l through 0 (Table 1.3).

In order to understand how electrons of many-electron atoms arrange themselves into the available orbitals it is necessary to define a fourth quantum number:

- The **spin quantum number**, m_s, can have two possible values, $+\frac{1}{2}$ or $-\frac{1}{2}$. These are interpreted as indicating the two opposite directions in which the electron can spin, $\uparrow$ and $\downarrow$.

Table 1.2 Values of principal and angular-momentum quantum numbers

N	ℓ
1	0 (1s)
2	0 (2s) 1 (2p)
3	0 (3s) 1 (3p) 2 (3d)
4	0 (4s) 1 (4p) 2 (4d) 3 (4f)

Table 1.3   Values of angular-momentum and magnetic quantum numbers

ℓ	Orbital	m_l	Representing
0	s	0	an s orbital
1	p	−1, 0, 1	3 equal-energy p orbitals
2	d	−2, −1, 0, 1, 2	5 equal-energy d orbitals
3	f	−3, −2, −1, 0, 1, 2, 3	7 equal-energy f orbitals

The total spin, S, of a number of electrons can be determined simply as the sum of the spin quantum numbers of the electrons involved and a state can be specified by its **spin multiplicity**:

$$S = \Sigma\, m_s$$
$$\text{Spin multiplicity} = 2S + 1$$

A ground-state helium atom has two paired electrons in the 1s orbital ($1s^2$). The electrons with paired spin occupy the lowest of the quantised orbitals shown below (the **Pauli exclusion principle** prohibits any two electrons within a given quantised orbital from having the same spin quantum number):

$$\text{Total spin } S = \frac{1}{2} - \frac{1}{2} = 0$$
$$\text{Spin multiplicity} = (2S + 1) = 1$$

This species is referred to as a **ground-state singlet** and is designated by S_0.

Electronic excitation can promote one of the electrons in the 1s orbital to an orbital of higher energy so that there is one electron in the 1s orbital and one electron in a higher-energy orbital. Such excitation results in the formation of an excited-state helium atom.

In the lowest excited-state helium atom there are two possible spin configurations:

1. The two electrons have opposite spins:

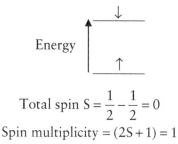

$$\text{Total spin } S = \frac{1}{2} - \frac{1}{2} = 0$$
$$\text{Spin multiplicity} = (2S + 1) = 1$$

A species such as this is referred to as an **excited singlet state**.
2. The two electrons have parallel spins:

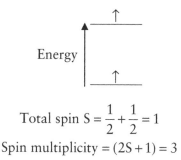

$$\text{Total spin } S = \frac{1}{2} + \frac{1}{2} = 1$$
$$\text{Spin multiplicity} = (2S + 1) = 3$$

In this case the species is referred to as an **excited triplet state**.

For a helium atom with two (paired) electrons in the 1s orbital in the ground state, the ground-state singlet is designated S_0, with S_1, S_2 ... being used to designate excited singlet states of increasing energy. Similarly T_1, T_2 ... are used to designate excited triplet states of increasing energy.

1.3 MODELLING MOLECULES: MOLECULAR ORBITALS

Some aspects of the bonding in molecules are explained by a model called **molecular orbital theory**. In an analogous manner to that used for atomic orbitals, the quantum mechanical model applied to molecules allows only certain energy states of an electron to exist. These quantised energy states are described by using specific wavefunctions called **molecular orbitals**.

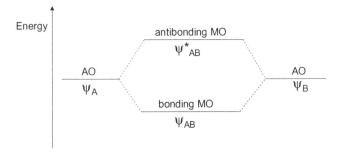

Figure 1.4 Formation of molecular orbitals (MO) by the interaction of two identical atomic orbitals (AO)

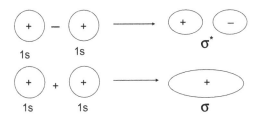

Figure 1.5 Boundary surfaces of σ (bonding) and σ^* (antibonding) molecular orbitals

In order to examine molecular orbitals at their simplest, we shall consider the case of diatomic molecules. The interaction of the wavefunctions of two identical atomic orbitals gives rise to wavefunctions of two distinct molecular orbitals (Figure 1.4).

The lower-energy **bonding molecular orbitals** result when atomic orbital wavefunctions enhance each other in the region of the nuclei. The atoms are held together by attraction between the nuclei and the electrons in the bonding molecular orbital, and $\psi_{AB} = \psi_A + \psi_B$.

Higher-energy **antibonding molecular orbitals** are formed when the atomic orbital wavefunctions cancel each other in the region of the nuclei, repelling electrons from the region where $\psi^*_{AB} = \psi_A - \psi_B$. The antibonding molecular orbital therefore represents a situation which tends to separate the atoms rather than bonding them together.

If the two atomic orbitals are s-orbitals then the resulting molecular orbitals are called σ (**sigma**) bonding molecular orbitals and σ^* (**sigma star**) antibonding molecular orbitals (Figure 1.5).

Molecular orbitals formed by mixing two parallel p-orbitals are called π (**pi**) if bonding and π^* (**pi star**) if antibonding (Figure 1.6).

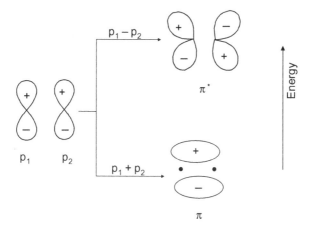

Figure 1.6 Boundary surfaces of π (bonding) and π^* (antibonding) molecular orbitals

The **phasing** of the molecular orbitals (shown as +/−) is a result of the wavefunctions describing the orbitals. + shows that the wavefunction is positive in a particular region in space, and − shows that the wavefunction is negative.

In addition to bonding and antibonding molecular orbitals it is necessary to consider **nonbonding molecular orbitals** (n). These orbitals contain lone pairs of electrons that are neither bonding nor antibonding and so play no part in bonding atoms together, being localised on just one atom. Nonbonding orbitals generally have a higher energy than bonding orbitals.

Figure 1.7 shows a simple representation of the relative quantised energy levels found in organic molecules.

The molecular orbital model can also be applied to complexes of the d-block elements. In octahedral complexes the d-orbitals of the metal are not degenerate, as they are in the free metal, because of the interaction between the ligand and metal orbitals. The five d-orbitals are split into three t_{2g} (nonbonding) and two e_g^* (antibonding) MOs; that is:

$$e_g^* \quad -- \quad (dx^2 - y^2 \, dz^2)$$

$$t_{2g} \quad --- \quad (dxy \; dxz \; dyz)$$

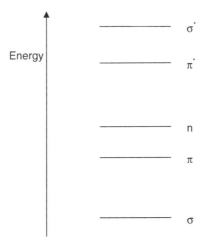

Figure 1.7 Schematic representation of molecular orbital energies in organic molecules

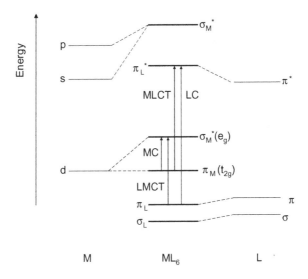

Figure 1.8 Molecular orbital diagram for an octahedral d-block metal complex ML_6. The vertical arrows indicate different types of electron transition that may be brought about by photon absorption

A representative molecular orbital diagram for an octahedral d-block metal complex ML_6 is shown in Figure 1.8. The MOs are classified as bonding (σ_L and π_L), nonbonding (π_M) and antibonding (σ_M^*, π_L^* and σ_M^*). The ground-state electronic configuration of an octahedral complex of a d^n metal has the σ_L and π_L MOs completely filled, while the n d-electrons are found in the π_M nonbonding MOs of t_{2g} symmetry and the σ_M^* antibonding MOs of e_g symmetry.

Photon absorption produces excited electron configurations by promotion of an electron from an occupied to a vacant MO. These electronic transitions are described as:

- Metal-centred (MC) transitions or d–d transitions between the nonbonding and antibonding metal-centred MOs. Such transitions are commonly found among the first-row d-block elements.
- Ligand-centred (LC) transitions between bonding and antibonding ligand-centred MOs. These transitions are expected for aromatic ligands with extended π- and π^*-orbitals.
- Ligand-to-metal charge transfer (LMCT) transitions between the bonding ligand-centred MOs and antibonding metal-centred MOs. Such transitions are found where a ligand is easily oxidised and the metal is easily reduced.
- Metal-to-ligand charge transfer (MLCT) transitions between the nonbonding metal-centred MOs and antibonding ligand-centred MOs. Such transitions are found where a metal is easily oxidised and the ligand is easily reduced.

1.4 MODELLING MOLECULES: ELECTRONIC STATES

The physical and chemical properties of molecules are determined by the electronic distribution within the molecules. For example, the ground-state methanal molecule, like other organic molecules, has a **closed shell structure**; that is, its occupied molecular orbitals all contain two paired electrons. Let us consider the possible electronic configurations that can occur in ground-state and excited-state methanal molecules. The molecular orbital diagram for methanal (formaldehyde), $H_2C=O$, is shown in Figure 1.9. The highest occupied molecular orbital (HOMO) is the nonbonding n molecular orbital localised on the O atom of the carbonyl group, and the lowest unoccupied molecular orbital (LUMO) is the antibonding π^* molecular orbital of the CO group.

Neglecting the filled low-energy σ-orbitals, the electronic configuration of the lowest electronic state (the ground state) is $\pi^2 n^2$. Photon absorption of the appropriate energy results in excited-state configurations by promotion of one electron from an occupied molecular orbital

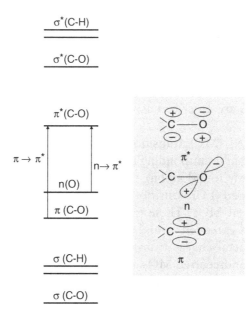

Figure 1.9 Molecular orbital diagram for methanal, showing the n → π* and π → π* transitions. The boundary surfaces of the π, n and π* molecular orbitals are also shown

to a vacant molecular orbital. The relatively low-energy n → π* (HOMO → LUMO) and π → π* electronic transitions lead to $\pi^2 n\pi^*$ and $\pi n^2 \pi^*$ excited configurations.

In terms of the spin multiplicity of the ground state and excited states of methanal, the ground state is a singlet state (S_0), with the excited states being either singlets (S_1, S_2, etc.) or triplets (T_1, T_2, etc.) (Figure 1.10).

Both the S_1 and T_1 excited states arise from the promotion of an electron from the n molecular orbital to the π* molecular orbital. They are referred to as $^1(n,\pi^*)$ and $^3(n,\pi^*)$ states, respectively. The S_2 and T_2 states arise from the promotion of an electron from the π molecular orbital to the π* molecular orbital and are referred to as $^1(\pi,\pi^*)$ and $^3(\pi,\pi^*)$ states, respectively. The **state diagram** for methanal is shown in Figure 1.11. With regard to the different spin states in molecules, the following ideas are important:

- An excited triplet state always has a lower energy than that of the corresponding excited singlet state. This is in line with **Hund's rule**: when two unpaired electrons occupy different orbitals, there is a

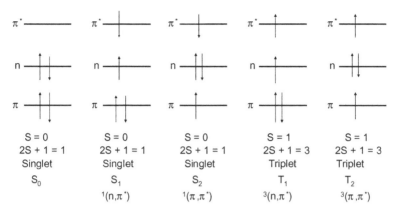

Figure 1.10 Configurations of the ground state and excited electronic states of methanal

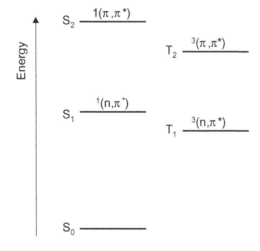

Figure 1.11 State diagram for methanal

minimum energy repulsion between the electrons when their spins are parallel.

- When electron excitation of a ground-state molecule, S_0, occurs, there is a tendency for the spin multiplicity to be retained. $S_0 \rightarrow T_1$ transitions are said to be **spin forbidden**; that is, they may be allowed but tend to be much less probable than $S_0 \rightarrow S_1$ transitions. Triplet excited states are generally accessed from the S_1 state only if the molecule or its environment favours the process of

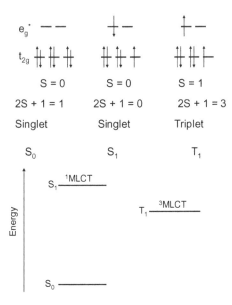

Figure 1.12 Electronic configurations and state diagram for the octahedral complex $Ru(bpy)_3^{2+}$. Only the lower-lying states are shown

intersystem crossing, whereby a molecule in the S_1 state is converted to a triplet state (see Section 2.6 and Chapter 5).

For octahedral complexes of Ru(III), and other d^6 ions, the σ_L and π_L MOs are fully occupied and the ground-state configuration is a closed shell because the HOMO π_M is fully occupied (t_{2g}^6). The ground state is a singlet state (S_0), with the excited states being either singlets or triplets (Figure 1.12).

However, in octahedral Cr(III) complexes, Cr^{3+} is a d^3 metal ion with three electrons in the HOMO orbitals π_M (t_{2g}^3). Thus the complexes of Cr(III) have an open-shell ground-state configuration, which on excitation produces quartet and doublet states (Figure 1.13).

1.5 LIGHT SOURCES USED IN PHOTOCHEMISTRY

Incandescent tungsten-filament lamps are good sources of visible light, with ultraviolet light down to 200 nm being generated by using a quartz tube. Below 200 nm, atmospheric oxygen absorbs ultraviolet light and so it is necessary to employ vacuum apparatus in order to work at these short wavelengths.

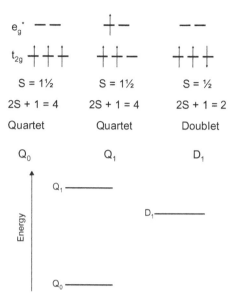

Figure 1.13 Electronic configurations and state diagram for a d^3 octahedral complex of Cr(III). Only the lower-lying states are shown

Discharge lamps contain xenon gas or mercury vapour through which an electric discharge is passed. At low pressure, the light they give out is emitted as a series of spectral lines characteristic of the element concerned. Discharge lamps have the ability to produce short pulses of light, and because of their high intensity at short wavelengths they are ideal for use as light pumping sources for lasers. Here, atoms are promoted to excited states, from which they can subsequently be stimulated to emit coherent monochromatic light.

Lasers are particularly important in photochemical research because their stimulated emission produces light which is monochromatic, coherent and very intense. Additionally, advances in laser technology have enabled lasers to deliver pulses of shorter and shorter duration such that today it is possible to produce pulses of the order of femto-seconds (1 fs = 10^{-15} s).

1.5.1 The Mercury Lamp

The mercury lamp has been the conventional light source used in photochemistry. The ground-state mercury atom, Hg, has two electrons in its highest occupied orbital, the 6s atomic orbital. Excited mercury atoms can therefore exist as either singlet or triplet spin states, rather like the case of helium discussed earlier.

Light emitted from a mercury lamp is caused by electronic transitions from higher-energy-level atomic orbitals to lower–energy-level atomic orbitals. The electronic transitions are subject to certain constraints known as **selection rules**:

- An electronic transition must involve a change in the orbital angular momentum quantum number ℓ such that $\Delta\ell = \pm 1$. Thus a 1s to 2p transition is allowed and a 1s to 3p transition is allowed, but a 1s to 2s or 1s to 3d transition is forbidden. This rule is sometimes called the **Laporte selection rule**.
- Where the electronic states involve singlets and triplets, transitions between the singlet and triplet states are forbidden by the **spin selection rule**, where there is no change in the spin multiplicity and thus $\Delta S = 0$. However, for electronic transitions in heavier atoms, such as mercury, there is a partial relaxing of the spin selection rule due to increased **spin–orbit coupling** in such atoms. Electrons in motion produce a magnetic field and spin–orbit coupling arises out of an interaction between the magnetic field due to the electron spin and the magnetic field due to the orbital motion of the electron. One of the consequences of this spin–orbit coupling is that the mixing of the states of different multiplicities occurs, so that the division into singlet and triplet states becomes less precise. The higher the atomic number, the greater the spin–orbit coupling and the less 'forbidden' the electronic transitions involving changes of multiplicity become.

The emission of a mercury lamp depends on its operating conditions. Low-pressure lamps have both the least intense emission and the fewest spectral output lines, the most prominent being a line at 254 nm. Medium-pressure lamps are brighter and produce a greater number of lines. High-pressure mercury lamps operating at high temperature and pressure have the most intense emission, which, because of extensive pressure-broadening, is continuous over a wide range of wavelengths, but the emission at 254 nm is absent because of the process of self-absorption.

1.5.2 Lasers

The word **laser** is an acronym for **light amplification by stimulated emission of radiation**.

Pulsed lasers emit pulses so short that even the fastest photoprocesses can be monitored, which would be impossible using conventional light sources.

The necessary conditions for the generation of laser light are:

1. A **laser medium** in a tube within the laser cavity, with a mirror at each end. A pump source must excite the atoms or molecules in the laser medium so that stimulated emission (Figure 1.3(b)) can occur. Lasers are classified by the nature of the medium, important ones being:
   - **Solid-state lasers**, such as the ruby laser, neodymium doped yttrium aluminium garnet (Nd-YAG) laser and the titanium doped sapphire laser.
   - **Gas lasers**, such as the helium-neon laser and the argon ion laser.
   - **Dye lasers**, in which the laser medium is a coloured fluorescent dye dissolved in a nonabsorbing solvent.

2. A **metastable state**, which is a relatively long-lived excited state with energy above the ground state. At thermal equilibrium the populations of the ground and metastable states are N_0 and N_1, respectively, where $N_0 \gg N_1$. If a system is to act as a laser then the incident photons must have a greater probability of bringing about stimulated emission than of being absorbed. This occurs when the rate of stimulated emission is greater than the rate of absorption; that is, when $N_1 \gg N_0$. This non-equilibrium condition is known as **population inversion**. Provided we can obtain a population inversion, light amplification can be achieved. We can now see why the metastable state needs to be relatively long-lived – it needs to exist for a sufficiently long time to allow a population inversion to build up.

   When a single photon of appropriate energy interacts with an excited-state atom, a second photon travelling in phase with and in the same direction as the initial photon will be emitted by stimulated emission. These two photons can cause further stimulated emission through interaction with two atoms, producing four photons by stimulated emission, and so on (Figure 1.14). Thus we have light **amplification**, which is also known as **gain**. The laser medium containing the atoms is sometimes known as the **gain medium**.

3. A **pumping mechanism**, in order to excite the atoms or molecules of the gain medium by supplying them with sufficient energy.

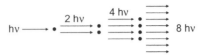

Figure 1.14 Light amplification resulting from stimulated emission (• represents an electronically-excited atom)

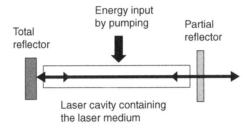

Figure 1.15 Schematic representation of a laser system

Depending on the type of laser used, this energy can come from a source of light, an electrical discharge or via chemical reactions.

4. An **optical resonance cavity** to sustain laser action. Even establishing population inversion and promoting stimulated emission is not enough to sustain laser action. In practice, photons need to be confined in the system to allow those created by stimulated emission to exceed all other mechanisms. This is achieved by the use of two mirrors bounding the laser medium, one totally reflecting, the other (in the direction of the output beam) partially reflecting (Figure 1.15).

1.5.2.1   How is a population inversion achieved in a laser?

At thermal equilibrium, according to the **Boltzmann distribution** (see Section 2.3), the population of atoms and molecules in the excited state can never exceed the population in the ground state for a simple two-level system.

However, laser action requires that N_1 be greater than N_0 in order to bring about the population inversion, but this can only be achieved when three or more energy levels are considered:

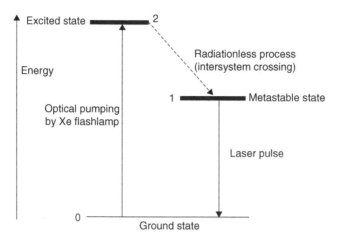

Figure 1.16 Transitions between energy levels in the ruby laser

- **The three-level ruby laser.** Ruby consists of Al_2O_3 containing traces of Cr^{3+}. A flash of white light from a xenon flashlamp excites the Cr^{3+} ions to state 2. They rapidly decay to state 1 by intersystem crossing. State 1 has a relatively long lifetime, allowing its population to build up so that it is greater than that of the ground state. In order to achieve population inversion, more than half of the ground-state Cr^{3+} ions must be excited (so that $N_1 > N_0$), which is not a particularly efficient process (Figure 1.16).
- **The four-level neodymium yttrium aluminium garnet (Nd-YAG) laser.** This solid-state laser consists of a rod of yttrium aluminium garnet doped with Nd^{3+} ions. The energy levels are shown in Figure 1.17. Optical pumping by a Xe flashlamp results in excitation of Nd^{3+} ions to level 3, followed by rapid radiationless decay to level 1. The lower laser level (level 2) has a very small population and so population inversion between levels 1 and 2 is easily achieved. The laser output (1064 nm) is in the infrared; **frequency-doubling** techniques allow it to be shifted into the visible and/or ultraviolet. Frequency-doubling, also known as **second harmonic generation**, involves absorption of two photons of frequency ν in a suitable crystal, such as KH_2PO_4, resulting in emission of a single photon of frequency 2ν. Frequency-doubling crystals can be used in series to produce other harmonics of the fundamental frequency of the laser.

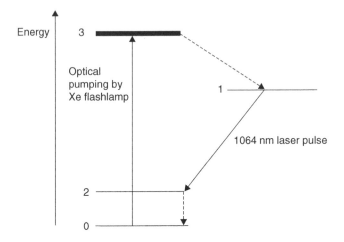

Figure 1.17   Transitions between energy levels in the neodymium laser

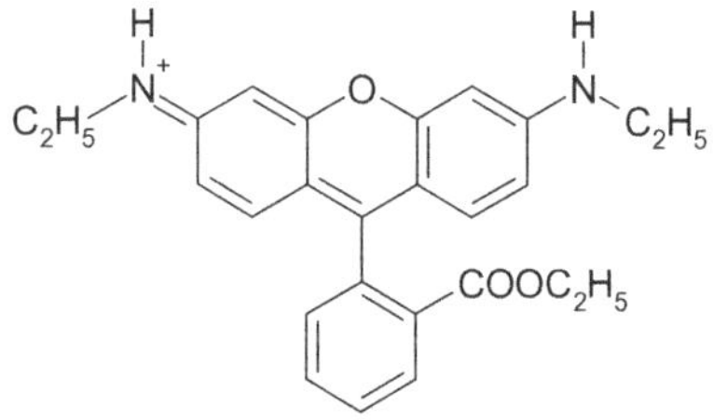

Figure 1.18   Rhodamine 6G

- **The four-level dye laser.** Whereas transitions between energy levels in atoms have a very narrow spread of energies, transitions between states of large molecules in solution are often very broad. This is the case for many solutions of dye molecules that have long conjugated chains and absorb in the visible region. The rhodamine 6G (Figure 1.18) laser allows the production of tuneable laser radiation where the output covers a broad spread of wavelengths (570–620 nm).

  Dye lasers are usually pumped by another laser, and a selection of around 25 dyes typically provides coverage of a wide region from 350 to 900 nm. Organic dye lasers are four-level lasers, even though only two electronic levels may be used (Figure 1.19).

  Vibrational relaxation (dotted arrows in Figure 1.19) is able to maintain a population inversion. The broad absorption and

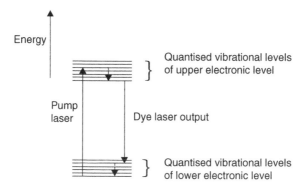

Figure 1.19   Energy levels and transitions in the rhodamine 6G dye laser

emission spectrum of rhodamine 6G, resulting from the broad upper and lower levels, enables the laser to be tuned to the required wavelength by use of a wavelength-selective device in the laser cavity.

1.5.2.2   Laser pulsing techniques

Some lasers produce a continuous-wave (CW) beam, where the timescale of the output cycle is of the same order as the time taken to remove photons from the system. CW lasers can be modified to produce a pulsed output, whereas other lasers are inherently pulsed due to the relative rates of the pumping and emission processes. For example, if the rate of decay from the upper laser level is greater than the rate of pumping then a population inversion cannot be maintained and pulsed operation occurs.

Pulsed lasers produce extremely short flashes of light, which means that photochemical events can be initiated very rapidly and subsequent physical and chemical events can be followed as they occur. In order to reduce the time of the laser pulse a number of techniques have been developed:

- Intense nanosecond pulses ($1\,\text{ns} = 10^{-9}\,\text{s}$) can be produced by **Q-switching**. A shutter is placed between the laser rod and one of the mirrors, thus inhibiting lasing. If the shutter is suddenly opened, the excitation is dumped in one huge burst. One type of shutter is the electro-optic **Pockels cell** (KH_2PO_4 crystal with a high applied potential).

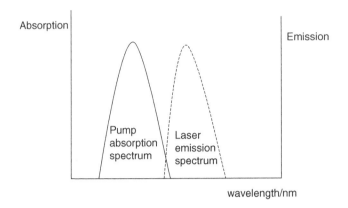

Figure 1.20   Absorption and emission spectra of Ti³⁺ ions in an alumina crystal

- Picosecond pulses ($1\,\mathrm{ps} = 10^{-12}\,\mathrm{s}$) are produced by **mode locking**, where the electronics of the instrumentation open and close the shutter at the round-trip frequency of the cavity.

In the last decade or so, the most remarkable advances in laser pulse generation have been carried out using the Ti^{3+} doped aluminium oxide laser, known as the **titanium sapphire laser** (Ti:sapphire). The absorption band of Ti^{3+} is in the blue-green spectral range, while the emission spectrum is shifted to higher wavelength, as shown in Figure 1.20. The maximum absorption occurs around 510 nm, in the green part of the spectrum, and so Ti:sapphire lasers are pumped by either the green line of an argon laser (514 nm) or the frequency-doubled (green) line of the Nd-YAG laser (532 nm). The broad absorption band allows tuning between 700 and 900 nm and the high peak output intensity enables the Ti:sapphire laser to be used as an efficient source for frequency doubling, giving 350–450 nm photons.

Ti:sapphire lasers can operate both in CW mode and in pulsed mode. There are two types of pulsed Ti:sapphire laser:

- One which produces pulses of the order of 10 ns and is usually pumped by a Q-switched Nd-YAG laser.
- A mode-locked Ti:sapphire laser, which can produce sub-ps pulses and is usually pumped by CW argon or Nd-YAG lasers. The shortest pulses coming from this laser type are exceptionally short, of the order of a few femtoseconds.

1.6 EFFICIENCY OF PHOTOCHEMICAL PROCESSES: QUANTUM YIELD

The preparation of integrated circuit chips relies on a photochemical process using **photoresists** (polymer coatings designed to change some physical property on exposure to light, thus providing a means of distinguishing between exposed and unexposed areas). The surface of the substrate is coated with a thin layer of photoresist material, which is then selectively irradiated, through a stencil, with ultraviolet light. The light causes a chemical change in the exposed region and, depending on the system used, it is possible to wash away either the exposed or the unexposed region selectively, using an appropriate solvent. The process is repeated many times to produce the final, complex circuit board. Because of the vast scale on which this process operates, it is important to understand the photochemistry, so as to maximise the efficiency of the reaction and reduce the considerable energy costs needed to power the light source. The **quantum yield** is a measure of how efficiently the absorbed photons are utilised.

1.6.1 Primary Quantum Yield (ϕ)

After photon absorption there are different ways in which the excited state may be deactivated, so not every excited molecule will form a primary product. The quantum yield for this primary process is given by:

$$\phi = \frac{\text{number of bonds broken in the primary step}}{\text{number of photons absorbed}}$$

Consider the photodissociation of propanone in Scheme 1.1.

$$(CH_3)_2CO \quad \xrightarrow[\lambda > 266 \text{ nm}]{h\nu} \quad (CH_3)_2CO^* \longrightarrow \bullet CH_3 + \bullet CH_3CO$$

$$\phi = 1$$

$$(CH_3)_2CO \quad \xrightarrow[\lambda < 193 \text{ nm}]{h\nu} \quad (CH_3)_2CO^* \longrightarrow 2\bullet CH_3 + CO$$

$$\phi = 2$$

Scheme 1.1

$$\text{HI} \xrightarrow[\lambda = 300 \text{ nm}]{h\nu} \text{H}\bullet + \text{I}\bullet \quad \text{Primary process}$$

$$\left.\begin{array}{l} \text{H}\bullet + \text{HI} \longrightarrow \text{H}_2 + \text{I}\bullet \\ \\ \text{I}\bullet + \text{I}\bullet \longrightarrow \text{I}_2 \end{array}\right\} \text{Secondary processes}$$

overall $2 \text{ HI} \longrightarrow \text{H}_2 + \text{I}_2$

$\phi = 2$

Scheme 1.2

Initiation $\text{Cl}_2 \xrightarrow{h\nu} 2\text{Cl}\bullet$

$$\text{Propagation} \left\{\begin{array}{l} \text{Cl}\bullet + \text{RH} \longrightarrow \text{HCl} + \text{R}\bullet \\ \\ \text{R}\bullet + \text{Cl}_2 \longrightarrow \text{RCl} + \text{Cl}\bullet \end{array}\right.$$

$$\text{Termination} \left\{\begin{array}{l} \text{Cl}\bullet + \text{Cl}\bullet \longrightarrow \text{Cl}_2 \\ \text{R}\bullet + \text{R}\bullet \longrightarrow \text{R}_2 \\ \text{R}\bullet + \text{Cl}\bullet \longrightarrow \text{RCl} \end{array}\right.$$

Scheme 1.3

The species $\cdot\text{CH}_3$ and $\cdot\text{CH}_3\text{CO}$ are **radicals**: species containing unpaired electrons. Radicals are formed by homolytic fission of a covalent bond, where the electron pair constituting the bond is redistributed such that one electron is transferred to each of the two atoms originally joined by the bond.

1.6.2   Overall Quantum Yield (Φ)

The overall quantum yield is the number of molecules of reactant, R, consumed per photon of light absorbed.

$$\Phi = \frac{\text{number of molecules of R consumed}}{\text{number of photons absorbed by R}}$$

According to the Stark–Einstein law, Φ should be equal to 1. However, if secondary reactions occur, Φ can be greater than 1.

Consider the photodissociation of hydrogen iodide in Scheme 1.2.

Φ can be very large for **chain reactions,** with the propagation reactions acting as an amplifier of the initial absorption step.

Irradiation of mixtures of hydrocarbons and chlorine at suitable wavelengths leads to chlorination of the organic molecule (Scheme 1.3). Reactions have overall quantum yields in excess of 10^6 ($>10^6$ propagation cycles for each termination step).

2

Light Absorption and Electronically-excited States

AIMS AND OBJECTIVES

After you have completed your study of all the components of Chapter 2, you should be able to:

- Explain the intensity pattern of absorption spectra and the occurrence or absence of vibrational fine structure.
- Determine the lowest energy transition of a simple organic molecule from the molar absorption coefficient and the structural formula.
- Show an understanding of the various selection rules giving rise to the allowed and forbidden electronic transitions.
- Explain the shifts on $n \rightarrow \pi^*$ and $\pi \rightarrow \pi^*$ transitions produced by substitution, conjugation and change of solvent polarity.

2.1 INTRODUCTION

All photochemical and photophysical processes are initiated by the absorption of a photon of visible or ultraviolet radiation leading to the formation of an electronically-excited state.

Principles and Applications of Photochemistry   Brian Wardle
© 2009 John Wiley & Sons, Ltd

$$R + h\nu \rightarrow R^*$$

For an effective interaction between the photon and the absorbing material:

- There should be a correspondence between the energy of the photon and the energy of a pair of electronic energy levels in the absorber.
- The strongest absorptions occur when the initial and final wavefunctions (ψ and ψ^*) most closely resemble one another.

2.2   THE BEER–LAMBERT LAW

The extent of absorption of light varies a great deal from one substance to another, with the probability of absorption being indicated by the **molar absorption coefficient** (ε). As light is absorbed, the intensity of light entering the substance, I_{in}, is greater than the intensity of the emerging light, I_{out}, and there is an exponential relationship between the relative absorption (I_{out}/I_{in}) and the concentration (c) and path length (l) of the absorbing substance:

$$I_{out}/I_{in} = 10^{-\varepsilon cl}$$

Taking logarithms to the base 10 gives us:

$$\log(I_{out}/I_{in}) = -\varepsilon cl$$

Thus:

$$\log(I_{in}/I_{out}) = \varepsilon cl$$

The left-hand-side quantity is the **absorbance, A,** and the linear relationship between absorbance, concentration and path length is known as the **Beer–Lambert law:**

$$A = \varepsilon cl$$

The Beer–Lambert law can generally be applied, except where very high-intensity light beams such as lasers are used. In such cases, a considerable proportion of the irradiated species will be in the excited state and not in the ground state.

The units of ε require some explanation here as they are generally expressed as non-SI units for historic reasons, having been used in spectroscopy for many years.

- Concentration, c, traditionally has units of moles per litre, mol l^{-1}.
- Path length, l, traditionally has units of centimetres, cm.
- A has no units since it is a logarithmic quantity.

Since A = εcl:

$$\varepsilon = A/cl$$

So the units of ε usually given are: $cm^{-1} \times (mol\ l^{-1})^{-1} = l\ mol^{-1}\ cm^{-1}$.

For a given substance, the molar absorption coefficient varies with the wavelength of the light used. A plot of ε (or $\log\varepsilon$) against wavelength (or wavenumber) is called the **absorption spectrum** of the substance (Figure 2.1). The principal use of absorption spectra from the photochemist's point of view is that they provide information as to what wavelength (λ_{max}) a compound has at its maximum value of the molar absorption coefficient (ε_{max}). Thus, irradiation of the compound at λ_{max} allows optimum photoexcitation of the compound to be carried out. In

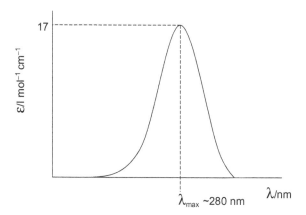

Figure 2.1   Absorption spectrum of propanone (acetone)

addition, the Beer–Lambert law is frequently used in the analytical determination of concentrations from absorbance measurements.

2.3   THE PHYSICAL BASIS OF LIGHT ABSORPTION BY MOLECULES

Light-absorbing molecules contain antennae groups known as **chromophores** or **chromophoric groups** which are responsible for the absorption of light. When the oscillating electromagnetic radiation encounters an appropriate chromophore, an electron in the chromophore can be promoted to a higher-energy excited state provided there is an energy correspondence between the photon and the pair of quantised electronic energy levels involved in the electronic transition. When this electronic transition occurs, the absorbing chromophore undergoes an **electric dipole transition** and the energy of the photon becomes part of the total energy of the excited-state molecule. The **transition dipole moment** lasts only for the duration of the transition and arises because of the process of electron displacement during the transition. The intensity of the resulting absorption is proportional to the square of the transition dipole moment.

In considering absorption of light by molecules, we have been principally concerned with transitions between electronic states. However, it is not possible to explain fully the effects of electronic excitation in molecules unless we also take into account the motions of the nuclei.

Now, the total energy of molecules is made up of electronic energy and energy due to nuclear motion (vibrational and rotational):

$$E_t = E_e + E_v + E_r$$

where the subscripts refer to the total energy, electronic energy, vibrational energy and rotational energy, respectively. Because of the large differences between electronic, vibrational and rotational energies, it is assumed that these can be treated separately. This assumption is known as the **Born–Oppenheimer approximation**.

The energy gap between electronic states is much greater than that between vibrational states, which in turn is much greater than that between rotational states. As a result, we are able to adequately describe the effects of electronic transitions within molecules by considering quantised electronic and vibrational states.

Absorption of ultraviolet and visible light by molecules results in electronic transitions in which changes in both electronic and vibrational states occur. Such transitions are called **vibronic transitions.**

At thermal equilibrium the population of any series of energy levels is described by the **Boltzmann distribution law.** If N_0 molecules are in the ground state then the number N_1 in any higher energy level is given by the equation:

$$\boxed{N_1/N_0 = \exp(-\Delta E/RT)}$$

where exp refers to the exponential function (e^x on calculators), ΔE is the energy difference between the two energy levels, R is the gas constant (which has a value of $8.314 \, J \, K^{-1} \, mol^{-1}$) and T is the absolute temperature.

Calculations based on the Boltzmann distribution law show that, at room temperature, most molecules will be in the $v = 0$ vibrational state of the electronic ground state and so absorption almost always occurs from $S_0(v = 0)$ (Figure 2.2).

Figure 2.3 shows the potential energy curve for a diatomic molecule, often referred to as a **Morse curve,** which models the way in which the potential energy of the molecule changes with its bond length.

At the points where the horizontal lines meet the Morse curve, the energy is wholly potential. In between, the energy is partly kinetic and partly potential.

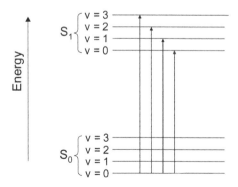

Figure 2.2 Schematic diagram of the electronic ground state and the first excited electronic state, with their associated quantised vibrational energy levels, for an organic molecule. The vertical arrows show vibronic transitions due to the absorption of photons

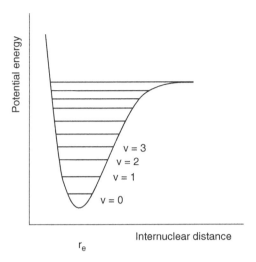

Figure 2.3 A Morse curve for a diatomic molecule, showing the quantised vibrational energy levels. The minimum on the curve represents the equilibrium bond distance, r_e

Each vibrational energy level has an associated wavefunction, the square of which relates to the most probable internuclear distance for a given vibrational quantum number, v. Figure 2.4 shows the vibrational probability function for a series of vibrational quantum numbers. For the v = 0 level, the square of the wavefunction shows that the molecule spends most of its time in the region of the equilibrium configuration. However, for an excited vibrational energy level, the magnitude of the ψ^2 function is greatest close to the turning points of the vibrational motion, which shows that the bond spends most of its time in the fully-compressed or fully-extended configuration.

Nuclei move much more slowly than the much-lighter electrons, so when a transition occurs from one electronic state to another, it takes place so rapidly that the nuclei of the vibrating molecule can be assumed to be fixed during the transition. This is called the **Franck–Condon principle**, and a consequence of it is that an electronic transition is represented by a vertical arrow such as that shown in Figure 2.5; that is, an electronic transition occurs within a 'stationary' nuclear framework. Thus the electronic transition accompanying the absorption of a photon is often referred to as a **vertical transition** or **Franck–Condon transition**.

Transitions between the vibrational levels in lower and upper electronic states will be most intense when the two states have similar

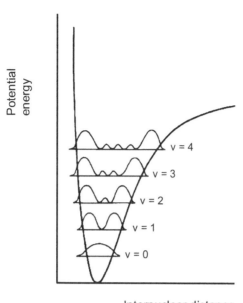

Figure 2.4 Vibrational probability functions for a series of vibrational quantum numbers. Note that for the higher v there is a greater probability of the molecule having a bond length at the two limits shown by the Morse curve. Notice also that for each value of v, there are v + 1 maxima

internuclear separations; that is, there will be a greater probability of an electronic transition when the ψ^2 functions for the upper and lower vibronic states have a greater overlap. This overlap, called the **Franck–Condon factor**, is shown in Figure 2.5.

2.4 ABSORPTION OF LIGHT BY ORGANIC MOLECULES

Figure 2.6 gives the UV–visible spectrum of a very dilute solution of anthracene (Figure 2.7) in benzene, which clearly shows small 'fingers' superimposed on a broader band (or envelope). These 'fingers' are called the **vibrational fine structure** and we can see that each 'finger' corresponds to a transition from the v = 0 of the ground electronic state to the v = 0, 1, 2, 3, etc. vibrational level of the excited electronic state.

The spectrum shows that many vibronic transitions are allowed, and that some are more probable than others; that is, the intensities of the

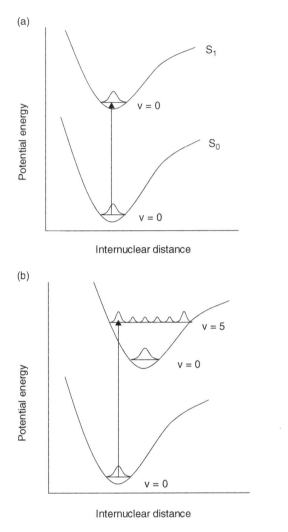

Figure 2.5   Electronic transitions with the greatest probability of absorption from $S_0(v = 0)$: (a) where both electronic states have similar geometries, shown by the minima of the curves being coincident ; (b) where the excited state has a larger internuclear distance than the ground state

different vibronic transitions vary. In the absorption spectrum of anthracene, the $v = 0 \rightarrow v = 0$ transition gives rise to the most intense absorption band because for this transition the overlap of the vibrational probability functions for $S_0(v = 0)$ and $S_1(v = 0)$ is greatest; that is, the Franck–Condon factor for this transition is greatest. The $v = 0 \rightarrow v = 0$ transition gives rise to the **0–0 band**.

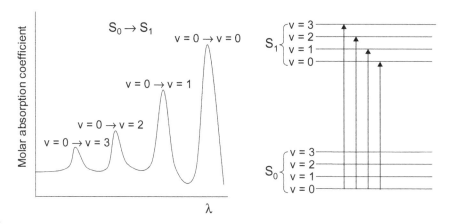

Figure 2.6 Absorption spectrum of a solution of anthracene in benzene, and the vibronic transitions responsible for the vibrational fine structure

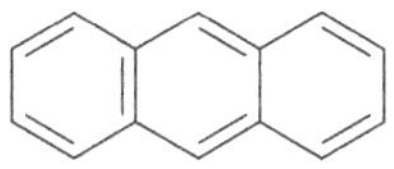

Figure 2.7 Anthracene

The absorption spectra of rigid hydrocarbons in nonpolar solvents may show vibrational fine structure, but absorption spectra of other organic molecules in solution tend to be broad, featureless bands with little or no vibrational structure (Figure 2.8). This is due to the very large number of vibrational levels in organic molecules and to blurring of any fine structure due to interaction between organic molecules and solvent molecules. The hypothetical spectrum shown in Figure 2.8 shows the vibrational structure hidden by the enveloping absorption spectrum, and the peak of the absorption curve does not correspond to the 0–0 band because the most probable vibronic transition here is the $0 \rightarrow 4$ transition.

The absorption bands in organic molecules result from transitions between molecular orbitals. The usual ordering of such molecular orbitals is shown in Figure 2.9.

Figure 2.9 shows that there are, in principle, six types of electronic transition, designated $\sigma \rightarrow \sigma^*$, $\sigma \rightarrow \pi^*$, $\pi \rightarrow \pi^*$, $\pi \rightarrow \sigma^*$, $n \rightarrow \sigma^*$ and $n \rightarrow \pi^*$. The $\sigma \rightarrow \sigma^*$ transitions correspond to absorption in the inaccessible far-ultraviolet, and both $\sigma \rightarrow \pi^*$ and $\pi \rightarrow \sigma^*$ are obscured by

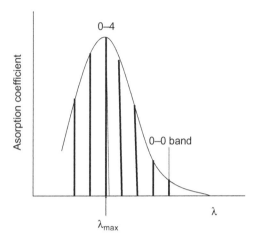

Figure 2.8 Broad, featureless absorption spectrum of the solution of an organic compound

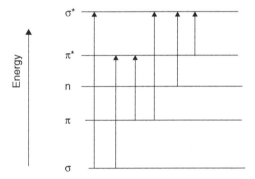

Figure 2.9 Generalised ordering of molecular orbital energies for organic molecules and electronic transitions brought about by excitation with light

the much stronger $\pi \rightarrow \pi^*$ absorptions. Of the possible electronic transitions, the ones we shall be most concerned with in molecular organic photochemistry are the $\pi \rightarrow \pi^*$ and $n \rightarrow \pi^*$ transitions, which produce (π,π^*) and (n,π^*) electronically-excited states, respectively.

In using the concept of molecular orbital theory to discuss the absorption of light by organic molecules, we concentrate on two molecular orbitals in particular. The highest occupied molecular orbital (HOMO) is the ground-state molecular orbital of highest energy with electrons in it and the lowest unoccupied molecular orbital (LUMO) is the

Table 2.1   Comparison of absorptions due to $\pi \to \pi^*$ and $n \to \pi^*$ electronic transitions

Absorptions due to $\pi \to \pi^*$ transitions	Absorptions due to $n \to \pi^*$ transitions
Occur at shorter wavelengths than do absorptions due to $n \to \pi^*$ transitions	Occur at longer wavelengths than do absorptions due to $\pi \to \pi^*$ transitions
Substitution moves the absorption to longer wavelength	Substitution moves the absorption to shorter wavelength
Relatively strong absorptions with ε_{max} values of ~10^3 to ~$10^5 \, l \, mol^{-1} cm^{-1}$	Relatively weak absorptions with ε_{max} values of ~1 to ~$10^2 \, l \, mol^{-1} cm^{-1}$
The absorption band occurs at longer wavelength in a polar solvent than in a nonpolar solvent (the absorption shows a **red shift** or **bathochromic shift**)	The absorption band occurs at shorter wavelength in a polar solvent than in a nonpolar solvent (the absorption shows a **blue shift** or **hypsochromic shift**)

ground-state molecular orbital of lowest energy with no electrons in it. Thus, the lowest energy transition in an organic molecule will be the HOMO $\to$ LUMO transition.

Absorptions due to $\pi \to \pi^*$ and $n \to \pi^*$ transitions differ from one another in several important respects, as shown in Table 2.1.

2.5   LINEARLY-CONJUGATED MOLECULES

When two carbon-to-carbon double bonds are present in a molecule, the effect on the electronic absorption spectrum depends on the distance between them. If the double bonds are conjugated then the first absorption band is encountered at considerably longer wavelength than that found for the molecule where the C=C bonds are isolated (not conjugated). When two π molecular orbitals are sufficiently close, overlap can occur, giving two delocalised π-orbitals, one of lower energy and one of higher energy. Likewise, the two π^*-orbitals give rise to two delocalised π^* molecular orbitals of different energies (Figure 2.10).

From Figure 2.10 it is obvious that the lowest-energy $\pi \to \pi^*$ transition in the conjugated diene ($\pi_2 \to \pi_3^*$) occurs at a lower energy than that for the isolated C=C bonds. Thus the effect of conjugation is to cause the first absorption band to move to a much longer wavelength (energy is inversely proportional to wavelength). The effect of further conjugation is to lower the energy of the $\pi \to \pi^*$ transition even further, as shown in Table 2.2.

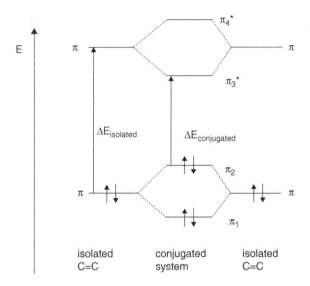

Figure 2.10   Interaction of two C=C units in a conjugated system

Table 2.2   Lowest energy absorption bands of conjugated polyenes $H(CH=CH)_nH$

n	λ_{max}/nm
2	217
3	268
4	304
5	334
6	364
7	390
8	410

Using propenal as an example, we can examine the conjugation between a C=C bond and a carbonyl group. The electronic absorption spectrum of propenal is shown in Figure 2.11.

The effect of conjugation on the π- and π^*-orbitals is analogous to the previous example with dienes but the n orbital remains virtually unchanged as a result of the conjugation. Figure 2.12 shows the effect of conjugation on the lowest-energy $\pi \rightarrow \pi^*$ and $n \rightarrow \pi^*$ transitions.

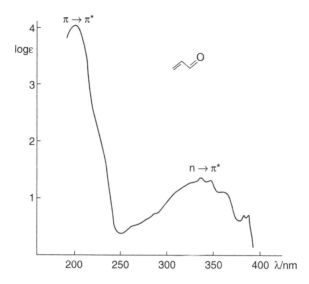

Figure 2.11   The electronic absorption spectrum of propenal

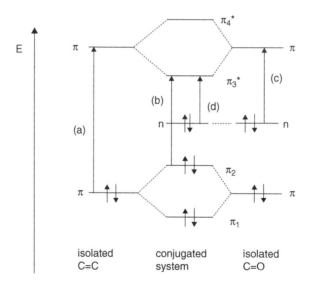

Figure 2.12   Interaction of a C=C bond and a carbonyl group in propenal. The $\pi \to \pi^*$ band, shown as transitions (a) and (b), occurs at lower energy (longer wavelength) as a result of conjugation. The $n \to \pi^*$, band shown as transitions (c) and (d), likewise occurs at longer wavelength as a result of conjugation

2.6   SOME SELECTION RULES

Transitions between energy levels in organic molecules are subject to certain constraints, referred to as **selection rules**.

1. **Spin selection rule:** An electronic transition takes place with no change in the total electron spin – that is, $\Delta S = 0$ – hence singlet $\leftrightarrow$ triplet transitions are forbidden or very weakly allowed. For example, the $S_0 \rightarrow T_1$ transition in anthracene has a molar absorption coefficient, ε_{max}, some 10^8 times less than that corresponding to the $S_0 \rightarrow S_1$ transition.

   The spin selection rule is derived from quantum mechanical calculations that do not take into account the interactions of electrons with other electrons or with nuclei in a molecule. **Spin–orbit coupling** results in the spin of an electron being affected by its orbital motion. As a consequence, a singlet state can be said to have some triplet character and a triplet state some singlet character, the result being that there is some mixing of the states and the spin selection rule is not rigidly applied. This is especially true for molecules containing atoms of high atomic mass (the so-called **heavy atom effect**).

   The heavy atom effect can show itself as the **internal heavy atom effect**, where incorporation of a heavy atom in a molecule will enhance $S_0 \rightarrow T_1$ absorption due to spin–orbit coupling. For example, 1-iodonaphthalene has a much stronger $S_0 \rightarrow T_1$ absorption than 1-chloronaphthalene

   The **external heavy atom effect** shows itself when a heavy atom is incorporated in a solvent molecule. For example, 1-chloronaphthalene has a much stronger $S_0 \rightarrow T_1$ absorption in iodoethane solution than in ethanol.

2. **Orbital symmetry selection rule:** According to the quantum theory, the intensity of absorption by molecules is explained by considering the wavefunctions of the initial and final states (ψ and ψ^*, respectively). An electronic transition will proceed most rapidly when ψ and ψ^* most closely resemble each other; that is, when the coupling between the initial and final states is strongest. Since the molar absorption coefficient, ε, is greatest when the electron transition is most probable (when the rate of absorption is greatest), the greatest values for ε also occur when the wavefunctions ψ and ψ^* most closely resemble each other. The weak absorption of the n $\rightarrow \pi^*$ transition compared to the $\pi \rightarrow \pi^*$ transition is a

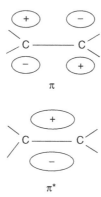

Figure 2.13 π and π^* molecular orbitals associated with the >C=C< chromophore. Both the π- and π^*-orbitals lie in the plane of the paper

consequence of the orbital symmetry selection rule. Transitions involving a large change in the region of space the electron occupies are forbidden. The orbital overlap between the ground state and excited state should be as large as possible for an allowed transition. π- and π^*-orbitals occupy the same regions of space, so overlap between them is large (Figure 2.13). The orbital overlap between n- and π^*-orbitals is very much smaller, as these orbitals lie perpendicular to each other (Figure 2.14).

The **phasing** of the molecular orbitals (shown as +/−) is a result of the wavefunctions describing the orbitals. + shows that the wavefunction is positive in a particular region in space and − shows that the wavefunction is negative.

Hence, according to the symmetry selection rule, $\pi \rightarrow \pi^*$ transitions are allowed but n $\rightarrow \pi^*$ transitions are forbidden. However, in practice the n $\rightarrow \pi^*$ transition is weakly allowed due to coupling of vibrational and electronic motions in the molecule (**vibronic coupling**). Vibronic coupling is a result of the breakdown of the Born–Oppenheimer approximation.

2.7 ABSORPTION OF LIGHT BY INORGANIC COMPLEXES

The metal complexes of the d-block elements absorb light due to electronic transitions occurring between d-orbitals of the metal species, or because of charge transfer within the complex.

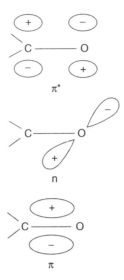

Figure 2.14 Molecular orbitals associated with the >C=O chromophore. The π- and π*-orbitals lie in the plane of the paper but the n-orbitals are perpendicular to the plane of the paper

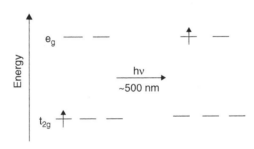

Figure 2.15 Absorption of light by $[Ti(H_2O)_6]^{2+}$ corresponding to a d → d transition

In a free d-block atom, all five d-orbitals are degenerate (all five have the same energy) but this is not the case in d-block metal complexes. In the octahedral complex $[Ti(H_2O)_6]^{2+}$, the five d-orbitals on the titanium are split into two sets: a triply-degenerate, lower-energy set (t_{2g}) and a doubly-degenerate, higher-energy set (e_g).

ε_{max} has a value of 6.1 at $\lambda_{max} \sim 500\,nm$ and corresponds to the promotion of the single d-electron in t_{2g} to e_g by absorption of visible light. Thus a solution of $[Ti(H_2O)_6]^{2+}$ has a pale purple colour (Figure 2.15).

As with organic compounds, the intensity of a transition is governed by selection rules. For inorganic complexes there are three selection rules to consider:

1. **Spin selection rule:** The spin selection rule, $\Delta S = 0$, specifies that there should be no change in the spin multiplicity. Weak spin-forbidden bands may occur when spin–orbit coupling is possible. Spin-forbidden transitions are more intense in complexes of heavy atoms as these lead to a larger spin–orbit coupling.
2. **Angular momentum selection rule:** $\Delta \ell = \pm 1$. Thus transitions that involve a change in angular momentum quantum number by 1 (i.e. $p \leftrightarrow d$ $d \leftrightarrow f$, for instance) are allowed. The important point here is that d–d transitions are **not** allowed.
3. **Laporte selection rule:** This is based on the symmetry of the complex. For complexes with a centre of symmetry, this forbids a transition between energy levels with the same symmetry with respect to the centre of inversion. So transitions such as $t_{2g} \rightarrow e_g$ are not allowed. The Laporte selection rule is not applied rigorously when ligands around the metal disrupt the perfect symmetry or when the molecule vibrates to remove the centre of symmetry. The interaction between electronic and vibrational modes is called vibronic coupling and means that d–d transitions are observed but are often weakly absorbing. Complexes with tetrahedral symmetry do not have a centre of inversion and so easier mixing of the levels is possible, giving more intense transitions.

Another way in which a metal complex may absorb light is by **charge-transfer transitions**, where transfer of an electron from the d-orbitals of the metal to the ligands or vice versa occurs. In such transitions, because the electron moves through a large distance the transition dipole moment will be large, resulting in an intense absorption. Charge-transfer transitions involve ligand- and metal-based levels, and are classified as **ligand-to-metal charge transfers** (LMCTs) and **metal-to-ligand charge transfers** (MLCTs). LMCTs can occur in all complexes, including ones with empty or completely filled d-shells. MLCT is seen in complexes in which the ligands have low-lying empty orbitals (e.g. CO and unsaturated ligands). An example of MLCT is in the ruthenium (II) trisbipyridyl complex, written as $Ru(bpy)_3^{2+}$, where, on photoexcitation, a d-electron is transferred from the ruthenium into the antibonding π^*-orbitals of a bipyridyl ligand and becomes delocalised over the extensive aromatic

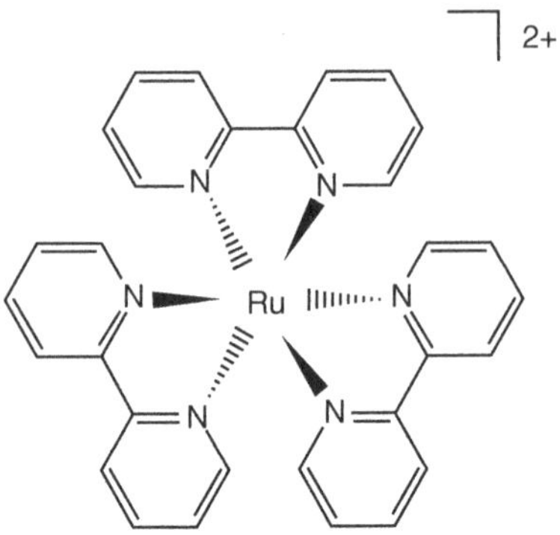

Figure 2.16   Structure of the octahedral ruthenium (II) trisbipyridyl complex. The orange colour of this complex results from metal-to-ligand charge-transfer (MLCT) transitions

ligand system. This results in the intense orange colour of the complex and a long excited-state lifetime (Figure 2.16).

The broad MLCT absorbance provides good overlap with the solar spectrum reaching earth and a vast research effort has been devoted to exploring the photophysical properties of $Ru(bpy)_3^{2+}$ and its derivatives. Much of this effort involves attempts to employ the ruthenium complexes as efficient light-absorbing and electron donors in both artificial photosynthesis (such as the splitting of water into hydrogen and oxygen) and solar cells (for converting sunlight into electrical energy).

3

The Physical Deactivation of Excited States

AIMS AND OBJECTIVES

After you have completed your study of all the components of Chapter 3, you should be able to:

- Explain the processes of absorption, radiative transitions and radiationless transitions in terms of Jablonski diagrams.
- Explain why for large molecules emission from higher excited states is rarely observed.
- Distinguish between the excited lifetime and radiative lifetime of S_1 and T_1 states.
- Understand the principles of time-correlated single-photon counting.

3.1 INTRODUCTION

Electronically-excited states of molecules are endowed with excess energy due to their formation by photon absorption. These excited states are short-lived, losing their excess energy within a very short period of time through a variety of deactivation processes (Figure 3.1) and returning to a ground-state configuration. If the excited molecule returns to its original ground state then the dissipative process is a

Principles and Applications of Photochemistry   Brian Wardle
© 2009 John Wiley & Sons, Ltd

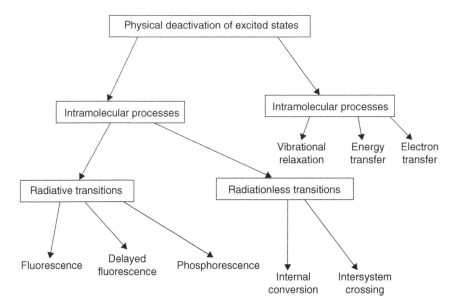

Figure 3.1 Physical deactivation of excited states of organic molecules

physical process, but if a new molecular species is formed then the dissipative process is accompanied by chemical change.

In this chapter we shall concentrate on an overview of the physical relaxation processes relating to organic molecules, along with a simple kinetic analysis of these processes. More detailed accounts of the processes themselves will be covered in subsequent chapters.

Physical relaxation processes may be classified as:

1. **Intramolecular processes**
   - **Radiative transitions** (Chapter 4), which involve the emission of electromagnetic radiation as the excited molecule relaxes to the ground state. Fluorescence and phosphorescence are known collectively as luminescence.
   - **Radiationless transitions** (Chapter 5), where no emission of electromagnetic radiation accompanies the deactivation process.
2. **Intermolecular processes**
   - **Vibrational relaxation**, where molecules having excess vibrational energy undergo rapid collision with one another other and with solvent molecules to produce molecules in the lowest vibrational level of a particular electronic energy level.

- **Energy transfer** (Chapter 6), where the electronically-excited state of one molecule (the donor) is deactivated to a lower electronic state by transferring energy to another molecule (the acceptor), which is itself promoted to a higher electronic state. The acceptor is known as a quencher and the donor is known as a sensitiser.
- **Electron transfer** (Chapter 6), considered as a photophysical process, involves a photoexcited donor molecule interacting with a ground-state acceptor molecule. An ion pair is formed, which may undergo back electron transfer, resulting in quenching of the excited donor.

3.2   JABLONSKI DIAGRAMS

The properties of excited states and their relaxation processes are conveniently represented by a **Jablonski diagram**, shown in Figure 3.2 and summarised in Table 3.1.

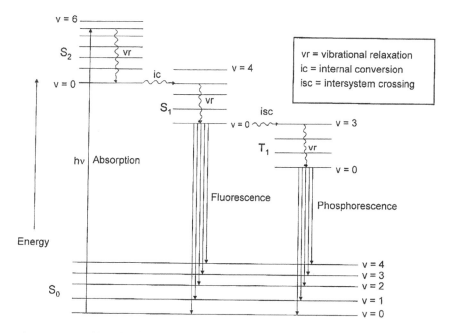

Figure 3.2   Jablonski diagram for an organic molecule, illustrating excited-state photophysical processes

Table 3.1 Summary of the photophysical processes shown in Figure 3.2

Relaxation Process	Details
Vibrational relaxation	Involves transitions between a vibrationally-excited state and the $v = 0$ state within a given electronic state when excited molecules collide with other species such as solvent molecules, e.g. $S_2(v = 3)$ ⤳ $S_2(v = 0)$. The excess vibrational energy is dissipated as heat.
Internal conversion	Involves radiationless transitions between vibronic states of the same total energy (isoenergetic states) and the same multiplicity. Internal conversion between excited states, e.g. S_2⤳ S_1 is much faster than internal conversion between S_1 and S_0
Intersystem crossing	Intramolecular spin-forbidden radiationless transitions between isoenergetic states of different multiplicity, e.g. S_1⤳ T_1
Fluorescence	Photon emission. Fluorescence involves a radiative transition between states of the same multiplicity (spin allowed), usually from the lowest vibrational level of the lowest excited singlet state, S_1. $$S_1(v = 0) \rightarrow S_0 + h\nu$$
Phosphorescence	Photon emission. Phosphorescence involves a spin-forbidden radiative transition between states of different multiplicity, usually from the lowest vibrational level of the lowest excited triplet state, T_1. $$T_1(v = 0) \rightarrow S_0 + h\nu$$

The Jablonski diagram shows:

- The electronic states of the molecule and their relative energies. Singlet electronic states are denoted by S_0, S_1, S_2, etc. and triplet electronic states as T_1, T_2, etc.
- Vibrational levels associated with each state are denoted as $v = 0$, $v = 1$, $v = 2$, etc. in order of increasing energy.
- Radiative transitions are drawn as straight arrows and radiationless transitions as wavy arrows.
- If an electronically-excited state is formed as a 'vibrationally-hot' excited molecule (with $v > 0$) then it will undergo vibrational relaxation within that electronic energy level until it reaches the $v = 0$ level. The vibrational relaxation within each electronically-excited state is drawn as a vertical wavy arrow.
- Radiationless transitions (internal conversion and intersystem crossing) between electronic states are isoenergetic processes and are drawn as wavy arrows from the $v = 0$ level of the initial state to a 'vibrationally-hot' ($v > 0$) level of the final state.

Notice that the energy difference between the excited states for each multiplicity is less than that between the ground state (S_0) and the first excited state. This results in the higher vibrational states of lower excited electronic states having similar energy to the lower vibrational states of higher excited electronic states. For example, as Figure 3.2 is drawn, the v = 3 level of S_1 has similar energy to the v = 0 level of S_2.

3.2.1   Vibrational Relaxation

An electronically-excited species is usually associated with an excess of vibrational energy in addition to its electronic energy, unless it is formed by a transition between the zero-point vibrational levels (v = 0) of the ground state and the excited state ($0 \rightarrow 0$ transition). Vibrational relaxation involves transitions between a vibrationally-excited state (v > 0) and the v = 0 state within a given electronic state when excited molecules collide with other species such as solvent molecules, for example S_2(v = 3) $\rightsquigarrow$ S_2(v = 0).

Typical timescales for the process are of the order of 10^{-13}–10^{-9} s in condensed phases, and the excess vibrational energy is dissipated as heat.

3.2.2   Internal Conversion

Relaxation from an upper excited electronic state such as S_2, S_3, etc. to a lower electronic excited state with the same multiplicity takes place rapidly by the radiationless process of internal conversion. Because the difference in energy of these upper excited states is relatively small, there is a high probability of the v = 0 level of, say, S_2 being very close in energy to a high vibrational level of S_1, allowing rapid energy transfer between the two electronic levels to occur. Because of the rapid rate of internal conversion between excited states, other radiative and nonradiative transitions do not generally occur from upper electronically-excited states as they are unable to compete with internal conversion.

Internal conversion involves intramolecular radiationless transitions between vibronic states of the same total energy (isoenergetic states) and the same multiplicity, for example S_2(v = 0) $\rightsquigarrow$ S_1(v = n) and T_2(v = 0) $\rightsquigarrow$ T_1(v = n). Typical timescales are of the order of 10^{-14}–10^{-11} s (internal conversion between excited states) and 10^{-9}–10^{-7} s (internal conversion between S_1 and S_0).

The much larger energy difference between S_1 and S_0 than between any successive excited states means that, generally speaking, internal conversion between S_1 and S_0 occurs more slowly than that between excited states. Therefore, irrespective of which upper excited state is initially produced by photon absorption, rapid internal conversion and vibrational relaxation processes mean that the excited-state molecule quickly relaxes to the $S_1(v_0)$ state from which fluorescence and intersystem crossing compete effectively with internal conversion from S_1. This is the basis of **Kasha's rule**, which states that because of the very rapid rate of deactivation to the lowest vibrational level of S_1 (or T_1), luminescence emission and chemical reaction by excited molecules will always originate from the lowest vibrational level of S_1 or T_1.

3.2.3   Intersystem Crossing

Intersystem crossing involves intramolecular spin-forbidden radiationless transitions between isoenergetic states of different multiplicity, for example $S_1(v = 0) \rightsquigarrow T_1(v = n)$. $S_1 \rightsquigarrow T_1$ intersystem crossing has a timescale of the order of 10^{-11}–10^{-8} s.

3.2.4   Fluorescence

Fluorescence involves a radiative transition (photon emission) between states of the same multiplicity (spin-allowed), usually from the lowest vibrational level of the lowest excited singlet state, $S_1(v = 0)$.

$$S_1 \rightarrow S_0 + h\nu$$

Typical timescales for fluorescence emission are of the order of 10^{-12}–10^{-6} s.

3.2.5   Phosphorescence

Phosphorescence is a spin-forbidden radiative transition between states of different multiplicity, usually from the lowest vibrational level of the lowest excited triplet state, $T_1(v = 0)$.

$$T_1 \rightarrow S_0 + h\nu$$

Typical timescales for photon emission by phosphorescence are of the order of 10^{-3}–10^{2} s.

3.3   EXCITED-STATE LIFETIMES

We saw in the last section that because of the rapid nature of vibrational relaxation and internal conversion between excited states an electronically-excited molecule will usually relax to the lowest vibrational level of the lowest excited singlet state. It is from the $S_1(v = 0)$ state that any subsequent photophysical or photochemical changes will generally occur (Kasha's rule).

3.3.1   Excited Singlet-state Lifetime

The competing intramolecular photophysical processes that can occur from $S_1(v_0)$ are fluorescence, intersystem crossing and internal conversion, with first-order rate constants of k_f, k_{isc} and k_{ic}, respectively (Figure 3.3).

Applying a standard treatment of first-order chemical kinetics, the rate of disappearance of the excited S_1 molecules, $^1J_{total}$, is given by:

$$^1J_{total} = -d[S_1]/dt = (k_f + k_{isc} + k_{ic})[S_1] = k_{total}[S_1]$$

(the – sign shows that the excited S_1 state decays with time; that is, its concentration decreases with time).

Solution of this equation gives the exponential decay form of the transient excited S_1 molecule:

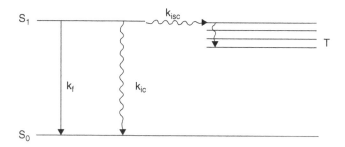

Figure 3.3   Competing photophysical processes occurring from S_1

$$[S_1]_t = [S_1]_0 \exp(-t/^1\tau)$$

where $[S_1]_0$ is the concentration of excited S_1 molecules at time t = 0 resulting from the initial exciting pulse, $[S_1]_t$ is the concentration of excited S_1 molecules at time t and $^1\tau$ is the **excited singlet-state lifetime** of the S_1 excited state.

When t = $^1\tau$:

$$[S_1]_t = [S]_0 \exp(-1) = [S]_0/e$$

Thus $^1\tau$ is given by the time for the concentration of S_1 to decrease to 1/e of its original value, where 1/e = 1/2.718 = 0.3679 ≈ 36.8%.

The excited singlet-state lifetime, $^1\tau$, is the time taken for the concentration of S_1 to decrease to 1/e of its initial value.

The technique of time-correlated single-photon counting (Figure 3.4) is used to measure an excited singlet-state lifetime, $^1\tau$. The sample is irradiated with a very short-duration light pulse (<<1 ns) to ensure any given molecule will only be excited once during the pulse. As soon as

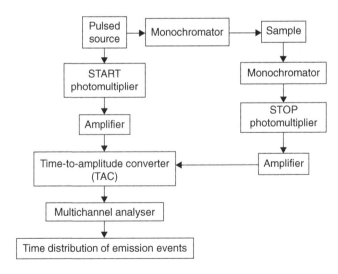

Figure 3.4 Schematic diagram of the principal components of a time-correlated single-photon counting apparatus

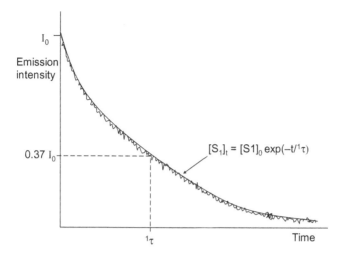

Figure 3.5   Time-resolved fluorescence decay measured by time-correlated single-photon counting, which involves counting the number of photons that arrive within a given time interval after excitation. The results are stored in a number of channels, each channel corresponding to a particular time interval. When displayed, the results are not continuous, but by using a large number of channels the output approximates to a continuous decay curve

the population of molecules is excited, the molecules randomly begin to relax to the ground state by fluorescence.

When the pulsed source produces its pulse of light to irradiate the sample, the START photomultiplier sends a signal to the TAC, which then linearly builds up a voltage until a signal is received from the STOP photomultiplier. When the STOP signal is received, the magnitude of the voltage (linearly proportional to the detection time) is measured and stored in a multichannel analyser. The multichannel analyser divides the voltage range into a sequence of several hundred channels, each channel building up the count of the number of times a certain voltage level is detected. The electronics of the system allows the process to be repeated millions of times a second, building up a histogram representing the time distribution of the emission events (Figure 3.5).

3.3.2   Excited Singlet-state Radiative Lifetime

The excited singlet-state radiative lifetime, $^1\tau_0$, of S_1 is the lifetime of S_1 in the absence of any radiationless transitions; that is, the only deactivation process is fluorescence. $^1\tau_0$ is the reciprocal of the rate constant for fluorescence, k_f:

$$\boxed{^1\tau_0 = 1/k_f}$$

Similarly, for the excited singlet-state lifetime:

$$\boxed{^1\tau = 1/(k_f + k_{isc} + k_{ic}) = 1/{}^1k_{total}}$$

where the sum of the rate constants for deactivation of the excited singlet state is given by:

$$^1k_{total} = (k_f + k_{isc} + k_{ic})$$

Since $^1k_{total}$ is greater than k_f, the observed excited singlet-state lifetime is less than the excited singlet-state radiative lifetime. $^1\tau$ only approaches $^1\tau_0$ as intersystem crossing and internal conversion from S_1 become much slower processes than fluorescence.

Now, the **fluorescence quantum yield**, ϕ_f, is the fraction of excited molecules that fluoresce. This is given by the rate of fluorescence, J_f, divided by the rate of absorption, J_{abs}:

$$\phi_f = J_f/J_{abs}$$

Under conditions of steady illumination, a steady state will be reached, where the rate of formation of excited molecules, $^1R^*$, is equal to the rate of deactivation by the intramolecular processes:

$$J_{abs} = {}^1J_{total}$$

Therefore:

$$\phi_f = J_f/J_{abs}$$
$$\phi_f = J_f/{}^1J_{total}$$
$$\phi_f = k_f[S_1]/{}^1k_{total}[S_1]$$
$$\boxed{\phi_f = k_f/{}^1k_{total}}$$

Now:

$$^1\tau_0 = 1/k_f \text{ and so } k_f = 1/{}^1\tau_0$$

Similarly:

$$^1\tau = 1/{}^1k_{total} \text{ and so } {}^1k_{total} = 1/{}^1\tau$$

Thus:

$$\boxed{\phi_f = {}^1\tau/{}^1\tau_0}$$

An order-of-magnitude estimate of the radiative lifetime of S_1 is given by:

$$\boxed{\tau_0 \approx 10^{-4}/\varepsilon_{max}}$$

where τ_0 has units of s and ε_{max} has units of l mol^{-1} cm^{-1}.

Thus, $\pi \rightarrow \pi^*$ transitions with ε_{max} of the order of 10^3–10^5l mol^{-1} cm^{-1} give $^1\tau_0$ of the order of ns–μs. For $n \rightarrow \pi^*$ transitions, ε_{max} has values of the order 10^0–10^2l mol^{-1} cm^{-1}, giving $^1\tau_0$ of the order of μs–ms.

3.3.3   Lifetimes of the T_1 Excited State

Similar considerations apply to the T_1 triplet state as to the S_1 singlet state. By analogy with the expressions for the lifetimes of S_1, the values for T_1 are given by:

$$\boxed{^3\tau_0 = 1/k_p}$$
$$^3\tau = 1/{}^1k_{tot} = 1/k_p + k_{isc}^{TS}$$

An order-of-magnitude estimate of the radiative lifetime of T_1 is given by:

$$\boxed{^3\tau_0 \approx 10^{-4}/\varepsilon_{max}}$$

where τ_0 has units of seconds and ε_{max} has units of l mol^{-1} cm^{-1}.

We saw in Chapter 2 that the process $S_0 + h\nu \rightarrow T_1$ is spin-forbidden.

The molar-absorption coefficients for such transitions will be very small and so T_1 states will have a longer lifetime than S_1 states. In general, $^3(\pi,\pi^*)$ states have longer radiative lifetimes $(1–10^2\,\text{s})$ than $^3(n,\pi^*)$ states $(10^{-4}–10^{-2}\,\text{s})$.

Because excited triplet states decay more slowly than excited singlet states, it is much easier to determine the excited triplet-state lifetime $^3\tau$ than $^1\tau$. Phosphorescence emission from a degassed sample at low temperature (77 K) lasts for longer than 1 ms and may even be several seconds. The molecules in the sample are irradiated with a short (~1 μs) flash and the decay of the phosphorescence signal is monitored using an oscilloscope. Any accompanying fluorescence signal will decay too rapidly to be observed. The excited triplet-state lifetime is obtained as the time taken for the emission intensity to fall to 1/e of its initial value.

4

Radiative Processes of Excited States

AIMS AND OBJECTIVES

After you have completed your study of all the components of Chapter 4, you should be able to:

- Describe the general features of a fluorescence and phosphorescence spectrum and explain their relationship to the absorption spectrum.
- Explain the basis of Kasha's rule and give an example of where this does not apply.
- Describe the essential features in the measurement of fluorescence and phosphorescence spectra, including any precautions necessary.
- Outline the essential features needed to determine a fluorescence quantum yield and a phosphorescence quantum yield.
- Assign differences in fluorescence quantum yield to differences in the electronic configuration of S1, substituent effects, molecular rigidity and the presence or absence of heavy atoms.
- Understand the principal mechanisms whereby certain molecules emit luminescence with the same spectral characteristics as fluorescence but with lifetimes more of the order of phosphorescence.

Principles and Applications of Photochemistry   Brian Wardle
© 2009 John Wiley & Sons, Ltd

4.1 INTRODUCTION

We saw in Section 3.1 that radiative processes of electronically-excited states can be categorised as fluorescence or phosphorescence, the distinction being based on the multiplicity of the two energy levels involved in the radiative processes. **Fluorescence** is the radiative transition from an excited state of the same spin multiplicity as the lower state. Fluorescence is strongly allowed and so occurs on relatively short timescales of the order of picoseconds to microseconds. If the electron spin multiplicity of the emitting and final states differs, the emission is known as **phosphorescence**. Phosphorescence is only weakly allowed (spin–orbit coupling promotes mixing of the singlet and triplet states) and thus phosphorescence emission is less intense and less rapid than fluorescence.

Radiative transitions may be considered as vertical transitions and may therefore be explained in terms of the Franck–Condon principle. The intensity of any vibrational fine structure associated with such transitions will, therefore, be related to the overlap between the square of the wavefunctions of the vibronic levels of the excited state and ground state. This overlap is maximised for the most probable electronic transition (the most intense band in the fluorescence spectrum). Figure 4.1 illustrates the quantum mechanical picture of the Franck–Condon principle applied to radiative transitions.

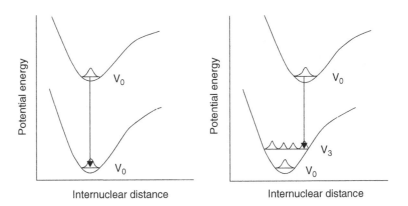

Figure 4.1 Most probable electronic transitions involved in radiative transitions where: (a) both electronic states have similar geometries; (b) the excited state and ground state have very different geometries

4.2 FLUORESCENCE AND FLUORESCENCE SPECTRA

Figure 4.2 shows the absorption and fluorescence emission spectra of a solution of anthracene in benzene (**1**) (Figure 4.3).

Mirror-image symmetry exists between absorption and fluorescence spectra of a solution of anthracene in benzene. This mirroring only occurs when the geometries of the ground state (S_0) and the first excited state (S_1) are similar.

The most noticeable features of the spectra, apart from the mirror-image relationship, are that:

- The 0–0 bands for absorption and fluorescence occur at almost the same wavelength.
- Fluorescence emission occurs at longer wavelengths (lower energy) than the 0–0 band, while absorption occurs at shorter wavelengths (higher energy) than the 0–0 band.
- The absorption spectrum shows a vibrational structure characteristic of the S_1 state whereas the fluorescence spectrum shows a vibrational structure characteristic of the S_0 state of anthracene.

Absorption occurs from $S_0(v = 0)$ and, because of rapid vibrational relaxation, fluorescence occurs from $S_1(v = 0)$. Fluorescence occurs at a lower energy (longer wavelength) than the exciting radiation because of

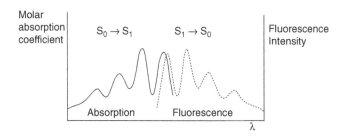

Figure 4.2 Absorption spectrum (continuous line) and fluorescence spectrum (dashed line) of anthracene in benzene

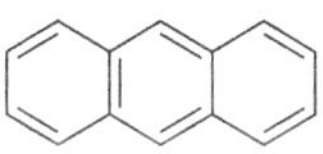

Figure 4.3 Anthracene (**1**)

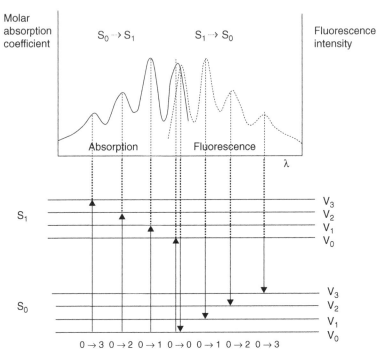

Figure 4.4 Energy-level diagram showing how the electronic and vibrational energy levels in the ground-state (S_0) and first excited-state (S_1) anthracene molecule are related to the absorption and fluorescence emission spectra

the vibrational energy that is transferred from the excited anthracene to its surroundings prior to fluorescence emission from $S_1(v = 0)$.

A simple energy-level diagram (Figure 4.4) shows that we would expect the 0–0 bands for fluorescence and absorption to occur at the same wavelength, since the energy changes (represented by the lengths of the arrows) are equal.

The 0–0 bands lie at slightly different wavelengths in absorption and emission. This separation results from energy loss to the solvent environment. For molecules in solution, the solvent cages surrounding the ground-state and excited-state molecules are different. Since electronic transitions occur at much faster rates than solvent-cage rearrangement, the energy changes involved in absorption and emission are different.

In rigid molecules, where the geometries of the S_0 and S_1 states are similar, there is a mirror-image relationship between the absorption spectrum and the fluorescence spectrum. This is due to the similarity of the energy spacing of the vibrational energy levels in the two states

(Figure 4.4). Because the vibrational energy level spacings in the S_0 and S_1 levels are similar, the 0–1 emission band is at the same energy below the 0–0 band as the 0–1 absorption band is above it, and so on for the other vibrational bands.

4.3   AN EXCEPTION TO KASHA'S RULE

According to Kasha's rule, fluorescence from organic compounds usually originates from the lowest vibrational level of the lowest excited singlet state (S_1). An exception to Kasha's rule is the hydrocarbon azulene (**2**) (Figure 4.5), which shows fluorescence from S_2.

This behaviour may be explained by considering that the azulene molecule has a relatively large S_2–S_1 gap, which is responsible for slowing down the normally rapid S_2 to S_1 internal conversion such that the fluorescence of azulene is due to the $S_2 \rightarrow S_0$ transition. The fluorescence emission spectrum of azulene is an approximate mirror image of the $S_0 \rightarrow S_2$ absorption spectrum (Figure 4.6).

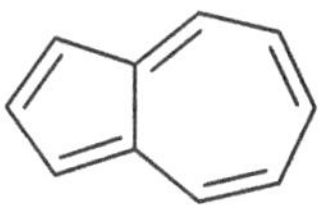

Figure 4.5   Azulene (**2**)

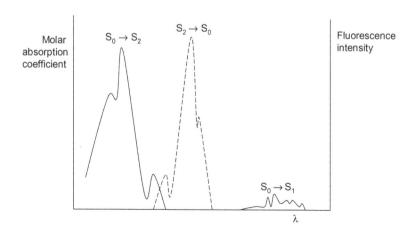

Figure 4.6   Absorption (continuous line) and fluorescence (dashed line) spectra of azulene
Adapted from N.J. Turro, *Modern Molecular Photochemistry* (1991). Copyright 1991 University Science Books

4.4   FLUORESCENCE QUANTUM YIELD

The ratio of the number of photons emitted by S_1 to the number of photons absorbed by S_0 is known as the fluorescence quantum yield, ϕ_f.

ϕ_F = Number of photons emitted by S_1/Number of photons absorbed by S_0

ϕ_f = Rate of emission of photons by S_1/Rate of absorption of photons by S_0

The fluorescence quantum yield of a compound may be determined by comparing the area under its fluorescence spectrum with the area under the fluorescence spectrum of a reference compound whose fluorescence quantum yield is known. The spectra of both compounds must be determined under the same conditions in very dilute solution using a spectrometer incorporating a 'corrected spectrum' capability, in order to overcome any variation in detector sensitivity with wavelength.

Because of Kasha's rule, the probability of an excited molecule ending up in the lowest vibrational energy level of S_1 is very high, irrespective of the energy of the exciting light used. Thus the fluorescence quantum yield is independent of the wavelength of the exciting light (**Vavilov's rule**).

When a molecule is in the $S_1(v = 0)$ state, fluorescence emission is only one of the several competing physical processes by which the molecule can return to the ground state. A molecule in $S_1(v = 0)$ can undergo fluorescence, intersystem crossing or internal conversion, which have rate quantum yields ϕ_f, ϕ_{isc} and ϕ_{ic}, respectively and:

$$\phi_f + \phi_{isc} + \phi_{ic} = 1$$

If the only process occurring from $S_1(v = 0)$ is fluorescence then ϕ_f will be equal to 1, whereas if no fluorescence occurs from $S_1(v = 0)$ then ϕ_f will be equal to 0. Thus the fluorescence quantum yield has values between 0 and 1.

In general, because of the relatively large energy gap between S_0 and S_1, ϕ_{ic} is much smaller than ϕ_f and ϕ_{isc}, which implies that $\phi_f + \phi_{isc} \approx 1$ (**Ermolev's rule**).

4.5 FACTORS CONTRIBUTING TO FLUORESCENCE BEHAVIOUR

A number of factors relating to the molecular structure of an organic compound and its environment have an effect on the fluorescence quantum yield, which, as we have seen, is dependent on the relative rates of the competing processes that may occur from $S_1(v = 0)$. These molecular factors are given below.

4.5.1 The Nature of S_1

Molecules in which S_1 is (π,π^*) tend to have a high fluorescence quantum yield. In general, $S_1(\pi,\pi^*)$ states have much shorter lifetimes (of the order of ns) than $S_1(n,\pi^*)$ states (of the order of μs) and so $S_1(\pi,\pi^*)$ states are more likely to undergo fluorescence emission before intersystem crossing can occur. Fluorescence from $S_1(n,\pi^*)$ is generally weak because not only is the excited state lifetime longer but the rate of intersystem crossing is faster. This faster rate of intersystem crossing is due to the fact that the singlet–triplet energy gap, E(S–T), is smaller for $S_1(n,\pi^*)$ than for $S_1(\pi,\pi^*)$. This may be explained by the enhanced spin–orbit coupling between $S_1(n,\pi^*)$ and the triplet state to which intersystem crossing occurs. Table 4.1 shows the importance of excited-state configuration in determining excited-state properties.

The net result is that for molecules in which S_1 is (π,π^*), both fluorescence and phosphorescence emission will be observed provided $\phi_f < 1$, but if S_1 is (n,π^*) then the quantum yield of phosphorescence will very likely be much greater than ϕ_f.

Table 4.1 Fluorescence properties of some representative compounds. The fluorescence quantum yields are measured in solution at room temperature

Compound	Structure	Nature of S_1	E(S–T)/kJ mol^{-1}	ϕ_f
naphthalene		(π,π^*)	132	0.19
benzophenone		(n,π^*)	21	1×10^{-6}

4.5.2　Molecular Rigidity

Molecular rigidity can be increased either by increasing the structural rigidity of the molecule (by preventing rotation or bending of bonds) or by increasing the rigidity of the medium (for example, by replacing a fluid solution at room temperature with a rigid glass made by freezing the fluid solution). Molecular rigidity favours efficient fluorescence emission, as is illustrated in Table 4.2.

The fluorescence quantum yield of trans-stilbene is 0.75 when measured in a rigid glass at 77 K, showing that the rigid medium results in more efficient fluorescence.

4.5.3　The Effect of Substituent Groups

Substituent groups have a marked effect on the fluorescence quantum yield of many compounds. Electron-donating groups such as -OH, $-NH_2$ and $-NR_2$ enhance the fluorescence efficiency, whereas electron-withdrawing groups such as -CHO, $-CO_2H$ and $-NO_2$ reduce the fluorescence quantum yield, as shown by naphthalene and its derivatives in Table 4.3.

4.5.4　The Heavy Atom Effect

The presence of so-called heavy atoms such as bromine or iodine in either the parent molecule (internal heavy atom effect) or the solvent

Table 4.2　The effect of molecular rigidity on fluorescence quantum yield measured in solution at room temperature

Compound	Structure	ϕ_f
trans-stilbene		0.05
5,10-dihydroindeno[2,1-a]indene		1.00
biphenyl		0.15
fluorene		0.66

Table 4.3 The effect of substituent groups on fluorescence efficiency of naphthalene and its derivatives. Fluorescence quantum yields measured in fluid solution at room temperature

Compound	ϕ_f
	0.19
NH$_2$	0.38
NO$_2$	0.0001

(external heavy atom effect) increases the probability of intersystem crossing by increasing the magnitude of the spin–orbit coupling. The heavy atom effect is illustrated in Tables 4.4 and 4.5.

4.6 MOLECULAR FLUORESCENCE IN ANALYTICAL CHEMISTRY

Measurement of fluorescence intensity can be used for quantitative analysis of fluorescent compounds where the intensity of fluorescence is proportional to the concentration of the compound. Because of their high sensitivity and selectivity, analytical techniques based on fluorescence detection are commonly used. If a target compound is fluorescent then direct detection of the fluorescence emitted is possible using a fluorimeter (Figure 4.7).

The monochromators are used to select the appropriate wavelengths for excitation and emission. The detector is placed at right angles to the incident light path, so that fluorescence, which is emitted in all directions, falls on the detector but no incident light is detected.

Numerous examples of the direct determination of compounds by fluorimetry have been reported in the literature. For example, the urine of patients with kidney disorders contains abnormally high levels of the

Table 4.4 The effect of the internal heavy atom effect on the fluorescence efficiency of naphthalene and its derivatives. Fluorescence quantum yields determined in solid solution at $77\,K$

Compound	ϕ_f
	0.55
Br	0.0016
I	0.0005

Table 4.5 The effect of the external heavy atom effect on the fluorescence efficiency of naphthalene. Fluorescence quantum yields determined in solid solution at $77\,K$

Solvent	ϕ_f
ethanol/methanol	0.55
1-bromopropane	0.13
1-iodopropane	0.03

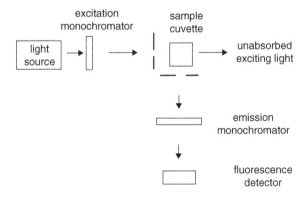

Figure 4.7 Basic components of a fluorimeter

hormones adrenaline, noradrenaline and dopamine. The three compounds of interest can be separated by high-performance liquid chromatography followed by fluorescence-intensity measurement at 310 nm (excitation wavelength 286 nm).

However, most molecules and ions show no fluorescence and so it is necessary to resort to other methods in these cases. **Derivatisation** methods can be used where the substance of interest is not fluorescent but may be converted into a fluorescent derivative by reaction with a suitable (nonfluorescent) reagent.

- The use of complexing reagents containing two functional groups is an effective method employed for metal ion determination by fluorescence measurement. For example, 8-hydroxyquinoline (3) (Figure 4.8) forms complexes with a large number of metal ions.

  (3) is not fluorescent, because its nonbonded electrons give rise to an (n,π^*) excited state on irradiation. However, when these electrons bind to Al^{3+} ions the complex (4) (Figure 4.9) that is formed is fluorescent.

  Complex (4) is fluorescent due to:
  - Formation of a ring, which increases the rigidity of the molecule.

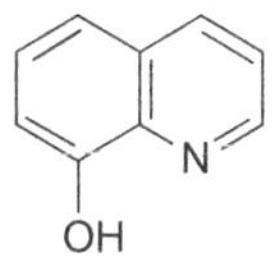

Figure 4.8   8-hydroxyquinoline (3)

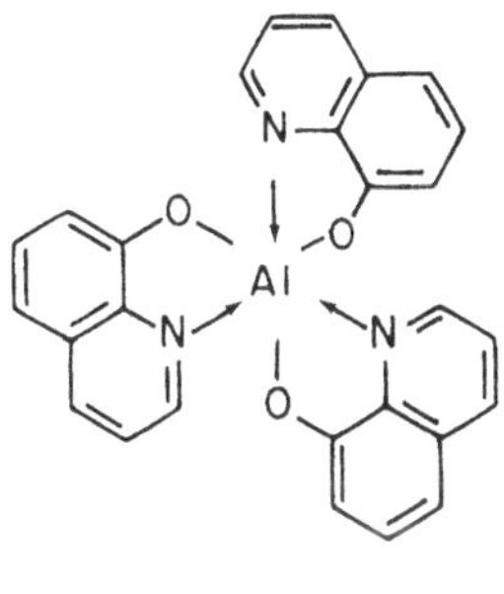

Figure 4.9   Complex (4)

- Electron-pair donation to the metal, removing the possibility of the low-lying n $\rightarrow$ π^* excited state, which would cause the reagent itself to be nonfluorescent.

Quenching methods, in which the fluorescence of a substance is reduced by the quenching action of an analyte, can also be used in fluorescence analysis. Such methods are particularly well suited to the analysis of gases.

By building scale models of aircraft equipped with many pressure sensors it is possible to investigate the effects of pressure variations across the model surface when high-speed wind tunnel trials are carried out. This method is very time-consuming and costly, so an alternative strategy involves the use of a paint containing an oxygen-sensitive fluorescent compound. The fluorescent lifetime of the pressure-sensitive paint becomes shorter as the oxygen pressure increases. In regions of high pressure, dissolved oxygen concentration in the paint increases and the fluorescence from the paint is quenched. Thus pressure images can be collected and related to previously collected pressure-calibration data.

4.7   PHOSPHORESCENCE

Phosphorescence arises as the result of a radiative transition between states of different multiplicity, $T_1 \rightarrow S_0$. Since the process is spin-forbidden, phosphorescence has a much smaller rate constant, k_p, than that for fluorescence, k_f:

$$k_f \left(\sim 10^6 - 10^9 \, s^{-1} \right) > k_p \left(10^{-2} - 10^4 \, s^{-1} \right)$$

Population of the triplet manifold by direct singlet–triplet absorption is a very inefficient process, being spin-forbidden. Instead, the triplet manifold is populated indirectly by excitation into the singlet manifold followed by intersystem crossing.

When the excited triplet state is populated, rapid vibrational relaxation and possibly internal conversion may occur (if intersystem crossing takes place to an excited triplet of greater energy than T_1). Thus the excited molecule will relax to the lowest vibrational level of the T_1 state, from where phosphorescence emission can occur in compliance with Kasha's rule.

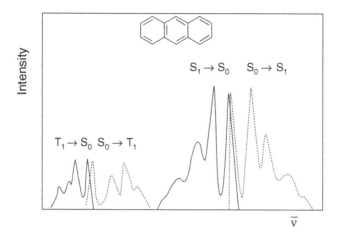

Figure 4.10 Absorption spectra (dashed line) and emission spectra (continuous line) of anthracene in solution in cyclohexane

Phosphorescence is spin-forbidden and thus phosphorescence emission is less intense (Figure 4.10) and less rapid than fluorescence.

Notice that in Figure 4.10 the intensity of absorption (molar absorption coefficient) and emission is plotted against wavenumber, which is proportional to energy. Because T_1 lies at lower energy than S_1, the phosphorescence spectrum is always found at lower wavenumbers (longer wavelengths) than the fluorescence spectrum.

Because $S_1 \rightarrow T_1$ absorption has a very small molar absorption coefficient, we would expect (because of the inverse relation between ε and τ_0) the T_1 state to have a much greater luminescent lifetime than the same molecules in the S_1 state. As a result of this longer lifetime, the T_1 state is particularly susceptible to quenching, such that phosphorescence in fluid solution is not readily observed as the T_1 state is quenched before emission can occur. This quenching in solution involves the diffusion together of either two T_1 molecules or the T_1 molecule and a dissolved oxygen molecule or some impurity molecule. In order to observe phosphorescence it is necessary to reduce or prevent the diffusion processes. The techniques most often used are:

- Freezing the solution by immersion in liquid nitrogen (77 K), ensuring that the solvent used does not result in the formation of an opaque solid but rather that a glassy solid is formed. The fluorescence is determined by using a **rotating-can phosphoroscope** (Figure 4.11). In addition to phosphorescence, fluorescence is also

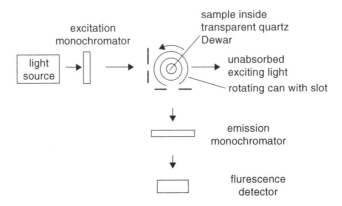

Figure 4.11   Schematic diagram of a rotating-can phosphoroscope

normally present. The two forms of luminescence are separated by exploiting the fact that T_1 states are much longer-lived than S_1 states and so phosphorescence persists long after the fluorescence has decayed. The rotation of the rotating can is set so that the path to the detector is blocked when the exciting light reaches the sample and open when the exciting light is blocked and the fluorescence has decayed.

• Room-temperature phosphorescence (RTP), dispersing the compound under investigation in a transparent polymer matrix such as Perspex

Because the triplet state is produced by intersystem crossing from S_1, not all absorbed photons result in the T_1 state that is able to emit phosphorescence.

The triplet quantum yield is given by:

$$\phi_T = k_{isc(ST)} / \left(k_{isc(ST)} + k_{ic} + k_f \right) = k_{isc(ST)}{}^1\tau$$

The fraction of triplet states that phosphoresce is given by the **phosphorescence quantum efficiency** (θ_p):

$$\theta_p = k_p / \left(k_p + \Sigma k_{nr(ST)} \right)$$

The **phosphorescence quantum yield** (θ_p) (the fraction of photons emitted from T_1 when S_1 is excited) is given by:

$$\phi_p = \phi_T \theta_p$$

The value of the phosphorescence quantum yield can be determined by measuring the total luminescence spectrum under steady irradiation. If the fluorescence quantum yield is known then the phosphorescence quantum yield may be found by comparing the relative areas under the two corrected spectra.

4.8   DELAYED FLUORESCENCE

In certain compounds a weak emission has been observed with the same spectral characteristics (wavelengths and relative intensities) as fluorescence, but with a lifetime more characteristic of phosphorescence. Two mechanisms are used to account for delayed fluorescence.

4.8.1   P-type Delayed Fluorescence (Triplet–Triplet Annihilation)

P-type delayed fluorescence is so called because it was first observed in pyrene. The fluorescence emission from a number of aromatic hydrocarbons shows two components with identical emission spectra. One component decays at the rate of normal fluorescence and the other has a lifetime approximately half that of phosphorescence. The implication of triplet species in the mechanism is given by the fact that the delayed emission can be induced by triplet sensitisers. The accepted mechanism is:

1. absorption: $S_0 + h\nu \rightarrow S_1$
2. intersystem crossing: $S_1 \rightarrow T_1$
3. triplet–triplet annihilation: $T_1 + T_1 \rightarrow X \rightarrow S_1 + S_0$
4. delayed fluorescence: $S_1 \rightarrow S_0 + h\nu$

It is the S_1 state produced by the triplet–triplet annihilation process that is responsible for the delayed fluorescence. Although it is emitted at the same rate as normal fluorescence, its decay is inhibited because it continues to be regenerated via step 3.

4.8.2   E-type Delayed Fluorescence (Thermally-activated Delayed Fluorescence)

E-type delayed fluorescence is so called because it was first observed in eosin.

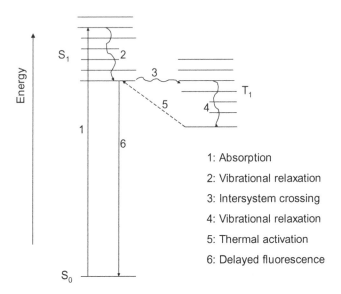

Figure 4.12 Jablonski diagram for the deactivation of a molecule by emission of E-type delayed fluorescence

The intensity of the delayed fluorescence emission from eosin decreases as the temperature is lowered and this indicates that an energy barrier is involved. Since the delayed fluorescence is spectrally identical to normal fluorescence, emission must occur from the lowest vibrational level of S_1. However, the fact that the lifetime is characteristic of phosphorescence implies that the excitation originates from T_1. The explanation of this requires a small S_1–T_1 energy gap, where T_1 is initially populated by intersystem crossing from S_1. T_1 to S_1 intersystem crossing then occurs by thermal activation.

The Jablonski diagram for thermally-activated delayed fluorescence is shown in Figure 4.12.

4.9 LANTHANIDE LUMINESCENCE

Trivalent lanthanide cations have luminescent properties which are used in a number of applications. The luminescence of the lanthanide ions is unique in that it is long-lasting (up to more than a millisecond) and consists of very sharp bands. Lanthanide emission, in contrast to other long-lived emission processes, is not particularly sensitive to quenching by oxygen because the 4f electrons found within the inner electron core

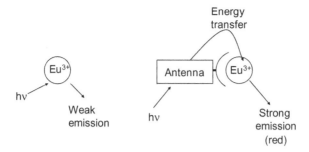

Figure 4.13 Indirect excitation of a europium (III) ion using an antenna chromophore

are not easily affected by external influences. The absorption and emission of lanthanide ions originates from f–f transitions, which are forbidden by spectral selection rules, thus implying that both absorption and emission are very weak. This problem can be overcome by complexing the lanthanide ion within a suitable organic ligand cage containing an antenna chromophore, which serves a number of purposes:

- The ligand prevents the quenching of the excited-state lanthanide by water.
- The ligand enhances the luminescence quantum yield.
- The ligand absorbs light efficiently and transfers the energy to the complexed lanthanide ion.

In 2000, most European countries switched from their traditional currencies to the euro. Lanthanide luminescence is used as a means of preventing counterfeit euro banknotes from passing into the money chain. Excitation of euro banknotes with ultraviolet light results in fluorescence in the red, green and blue regions due to complexes of europium (Eu^{3+}), terbium (Tb^{3+}) and thulium (Tm^{3+}), respectively, that are present in the banknotes.

The nature of the emission by these three lanthanide ions is phosphorescence, since the emission of light is accompanied by a change in spin multiplicity. For example, the emission by the Eu^{3+} cation involves a change in the spin multiplicity from 5 to 7 on going from the excited state to the ground state ($^{5}Eu^{*} \rightarrow {}^{7}Eu$).

The complexing ligand of the lanthanide cation also provides a way for the luminescent lanthanide complex to be linked to a particular target molecule, thus acting as a luminescent probe for the target molecule. Lanthanide complexes have been designed so that a strong luminescence emission is brought about in response to a particular molecular recognition event, such as protein folding or protein–DNA interaction. This is done by controlled arrangement of different ligands around the lanthanide ion. A nonabsorbing ligand (NAL) forms a stable complex with the lanthanide cation bonded to one molecular site, while a light harvesting centre (LHC) is bound to another molecular site. If the two sites interact then the LHC absorbs light and transfers its energy to the lanthanide, which then emits a characteristic visible phosphorescence. The lanthanide-NAL and LHC are placed in specific protein sites and only when these sites come together does the phosphorescence emission occur.

5

Intramolecular Radiationless Transitions of Excited States

AIMS AND OBJECTIVES

After you have completed your study of all the components of Chapter 5, you should be able to:

- Describe the general features of internal conversion and intersystem crossing.
- Understand that intermolecular radiationless transitions of excited states are caused by a breakdown of the Born–Oppenheimer approximation.
- Understand the importance of the overlap of vibrational probability functions and the energy gap law in determining the rate of internal conversion and intersystem crossing.
- Explain the effects of deuteration and heavy atoms on the relative rates of radiationless processes.
- Outline the basis of El-Sayed's rules and be able to use these to explain the differences in luminescence properties between aliphatic and aromatic carbonyl compounds.

5.1 INTRODUCTION

Radiative and radiationless (nonradiative) transitions may be pictured as competing vertical and horizontal crossings, respectively, between the

Principles and Applications of Photochemistry   Brian Wardle
© 2009 John Wiley & Sons, Ltd

v = 0 vibrational energy level of an upper electronic state (i) and a set of closely-spaced vibrational levels on a lower electronic energy surface (f_1 or f_2).

In the radiative transition shown, most of the energy is removed from the system by photon emission, whereas for the radiationless transition the sum of the electronic energy and vibrational energy is constant and energy is subsequently removed from the system by vibrational relaxation to v = 0 of f_2, with the solvent acting as an energy sink.

The relative magnitude of the spacing of the quantised electronic and vibrational states shows that the energy difference between the electronic states is much greater than that between the vibrational states. This generally allows us to treat the electronic and vibrational energy quite separately from each other (the Born–Oppenheimer approximation). In reality, this separation of electronic and vibrational energies is not exactly correct. This is because the molecular vibrations do affect the electronic energy of the molecule and so both the vibrational and electronic energies need to be taken into account when considering a radiationless transition from one state to another. The occurrence of radiationless (nonradiative) transitions is due to the breakdown of the Born–Oppenheimer approximation. As shown in Figure 5.1, a radiative transition occurs between vibronic states which differ in energy, whereas a radiationless transition occurs between vibronic states of the same energy. It is the linking of electronic and vibrational energies, considered together, that determines the efficiency (rate) of radiationless transitions.

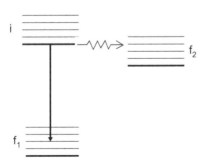

Figure 5.1 Competing radiative (vertical straight arrow) and radiationless (horizontal wavy arrow) processes between initial (i) and final (f) electronic states

5.2 THE ENERGY GAP LAW

The probability of intramolecular energy transfer between two electronic states is inversely proportional to the energy gap, ΔE, between the two states. The value of the rate constant for radiationless transitions decreases with the size of the energy gap between the initial and final electronic states involved. This law readily provides us with a simple explanation of Kasha's rule and Vavilov's rule.

The rate of internal conversion between electronic states is determined by the magnitude of the energy gap between these states. The energy gaps between upper excited states (S_4, S_3, S_2) are relatively small compared to the gap between the lowest excited state and the ground state, and so the internal conversion between them will be rapid. Thus fluorescence is unable to compete with internal conversion from upper excited states. The electronic energy gap between S_1 and S_0 is much larger and so fluorescence ($S_1 \rightarrow S_0$) is able to compete with $S_1(v = 0)$ $\rightsquigarrow S_0(v = n)$ internal conversion.

The efficiency of intersystem crossing is determined by the size of the singlet–triplet energy gap (singlet–triplet splitting), ΔE_{ST}. Now, $\Delta E_{ST}(n,\pi^*) < \Delta E_{ST}(\pi,\pi^*)$, with $\Delta E_{ST}(n,\pi^*)$ being $< 60\,kJ\,mol^{-1}$ and $\Delta E_{ST}(\pi,\pi^*)$ being $> 60\,kJ\,mol^{-1}$. Therefore intersystem crossing for an (n,π^*) excited state will be much faster than intersystem crossing for a (π,π^*) excited state.

5.3 THE FRANCK–CONDON FACTOR

In Chapters 2 and 4, the Franck–Condon factor was used to account for the efficiency of electronic transitions resulting in absorption and radiative transitions. The efficiency of the transitions was envisaged as being related to the extent of overlap between the squares of the vibrational wave functions, ψ^2, of the initial and final states. In a horizontal radiationless transition, the extent of overlap of the ψ^2 functions of the initial and final states is the primary factor controlling the rate of internal conversion and intersystem crossing.

The variations in efficiency (rate) of radiationless transitions result from differences in the Franck–Condon factor, visualised by superimposing the vibrational wavefunctions, ψ (or ψ^2 – the probability distributions), of the initial and final states. We will consider three cases illustrated in Figure 5.2.

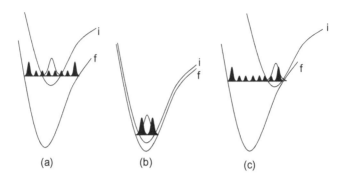

Figure 5.2 Overlap of ψ^2 functions for a radiationless transition between the initial (i) and final (f) electronic states
Adapted from C.E. Wayne and R.P. Wayne, *Photochemistry*, Oxford Chemistry Primers 39 (Oxford University Press, 1996)

5.3.1   Case (A): Both Electronic States Have a Similar Geometry, with a Large Energy Separation between the States

In Figure 5.2(a), both electronic states have similar geometries, shown by the nested curves with their minima being coincident. Their electronic energy separation is large, with the $v = 0$ vibrational level of the initial electronic state being close to the $v = 7$ vibrational level of the final electronic state. There is very little overlap between the isoenergetic ψ^2 functions and so the rate of radiationless transfer will be slow.

With large rigid aromatic molecules there is virtually no change in molecular geometry on excitation. The overlap of ψ^2 functions in Figure 5.2(a) shows that in such cases the rate of internal conversion from S_1 to S_0 will be very slow. Thus, in such cases, fluorescence is able to compete favourably with radiationless transfer and hence these rigid systems tend to fluoresce strongly.

5.3.2   Case (B): Both Electronic States Have a Similar Geometry, with a Small Energy Separation between the States

In Figure 5.2(b), both electronic states have similar geometries and their energy separation is small, with the $v = 0$ vibrational level of the initial electronic state being close to the $v = 1$ vibrational level of the final electronic state. There is a good overlap between the isoenergetic ψ^2 functions, leading to rapid radiationless transfer.

5.3.3 Case (C): The Electronic States Have Different Geometries, with a Large Energy Separation between the States

In the third situation, shown in Figure 5.2(c), the energy gap is large and the initial and final electronic states have very different geometries (shown by the minima corresponding to different internuclear distances). Even though the electronic energy gap is large, there is significant overlap between the isoenergetic ψ^2 functions of the v = 0 vibrational level of the initial electronic state and the v = 7 vibrational level of the final electronic state. Thus the rate of radiationless transfer is fast.

In rigid molecules that have nested potential energy curves for the ground and excited states, radiative processes are favoured when there is no significant overlap of the vibrational wavefunctions and internal conversion and intersystem crossing are weak. In nonrigid molecules, any significant geometry differences between the initial and final electronic states favour internal conversion and intersystem crossing and weaken the radiative emission.

Thus, the efficiency (rate) of internal conversion and intersystem crossing depends on both electronic and vibrational factors:

- Electronic factors describe the probability of transitions between one electronic state and another.
- Vibrational factors involve the operation of the vibrational wavefunction overlap with the radiationless transitions.

For aromatic hydrocarbons, the rate of radiationless transitions depends largely on the frequency of the C-H vibrations. These C-H vibrations are of high frequency and are widely spaced. If all the hydrogen atoms in a molecule are replaced by deuterium then the C-D vibrational energy levels are closer together than the C-H vibrational energy levels. Because of this smaller vibrational energy gap, the vibrational overlap will involve levels with a higher vibrational quantum number, such as that illustrated in Figure 5.3, and the Franck–Condon factor will be smaller.

Considering the deuterium effect on naphthalene, $C_{10}H_8$, in which all the hydrogen atoms are replaced by deuterium, the Franck–Condon factor is decreased and consequently the rate of internal conversion

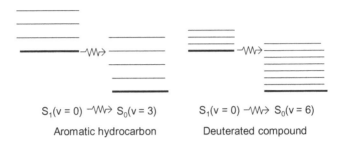

$S_1(v = 0)$ ⤳ $S_0(v = 3)$ $S_1(v = 0)$ ⤳ $S_0(v = 6)$

Aromatic hydrocarbon   Deuterated compound

Figure 5.3 The effect of energy gap in vibrational levels on S_1 ⤳ S_0 internal conversion. Decreasing the vibrational energy gap leads to a radiationless transition in which the ψ^2 overlap and Franck–Condon factor are reduced and the rate of internal conversion should be decreased

Table 5.1 Effect of deuteration on intersystem crossing rates (k_{TS}) and triplet-state lifetimes

Compound	k_{TS}/s^{-1}	$^3\tau/s$
naphthalene	0.39 (all H) 0.02 (all D)	2.4 (all H) 19.0 (all D)
anthracene	22.0 (all H) 7.1 (all D)	0.045 (all H) 0.14 (all D)
pyrene	2.0 (all H) 0.29 (all D)	0.5 (all H) 3.2 (all D)

should be decreased and the fluorescence quantum yield increased. Experimentally it is found that the fluorescence quantum yields for both $C_{10}H_8$ and $C_{10}D_8$ are the same, and this absence of a deuterium effect on fluorescence quantum yield indicates that S_1 ⤳ S_0 internal conversion for many molecules is of negligible importance.

Further proof of the importance of Franck–Condon factors is shown by the dramatically increased triplet-state lifetimes of aromatic hydrocarbons that have been deuterated. The effect of this deuteration is to decrease the rate of T_1 ⤳ S_0 intersystem crossing, which is accompanied by a corresponding increase in triplet-state lifetime (Table 5.1).

5.4 HEAVY ATOM EFFECTS ON INTERSYSTEM CROSSING

Intersystem crossing is a spin-forbidden process between states of different multiplicity, so the magnitude of the spin–orbit coupling is important in controlling the rate of intersystem crossing. Transitions between

Table 5.2 The effect of heavy atoms on transitions between states in rigid solution (77 K)

Molecule	k_{ST}/s^{-1}	k_{TS}/s^{-1}
naphthalene	10^6	10^{-1}
1-chloronaphthalene	10^8	10
1-bromonaphthalene	10^9	50

states of different multiplicity can take place if the triplet has some singlet character and if the singlet has some triplet character. This mixing of singlet and triplet character is brought about by spin–orbit coupling. It was shown in Section 4.5 that the presence of so-called heavy atoms in a molecule enhances the rate of intersystem crossing by increasing the spin–orbit coupling. Table 5.2 shows the heavy atom effect on S $\rightsquigarrow$ T and T $\rightsquigarrow$ S transitions in naphthalene and its derivatives.

5.5 EL-SAYED'S SELECTION RULES FOR INTERSYSTEM CROSSING

The extent to which a given molecule will undergo luminescence emission depends on the relative values of the rate constants of the competing processes occurring from the S_1 and T_1 states, as shown in Figure 5.4. Internal conversion between S_1 and S_0 and intersystem crossing between T_1 and S_0 are neglected due to the large energy gaps involved.

Thus S_1 and T_1 can relax to the ground state only by processes involving luminescence or chemical reaction.

The quantum yields of fluorescence and phosphorescence are given by:

$$\phi_f = k_f / (k_f + {}^1k_{isc} + {}^1k_r)$$
$$\phi_p = [k_p / (k_p + {}^3k_r)] \times [{}^1k_{isc} / ({}^1k_{isc} + k_f + {}^3k_r)]$$

Thus it is apparent that fluorescence is observed only when $k_f > {}^1k_{isc} + {}^1k_r$ and phosphorescence is observed only when both $k_p > {}^3k_r$ and ${}^1k_{isc} > k_f + {}^3k_r$.

Also, since the term for ${}^1k_{isc}$ occurs in the expressions for both ϕ_f and ϕ_p, the magnitude of ${}^1k_{isc}$ is important in determining the outcome of the process:

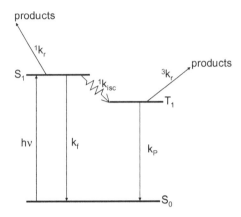

Figure 5.4 Competing photophysical and photochemical processes occurring from S_1 and T_1

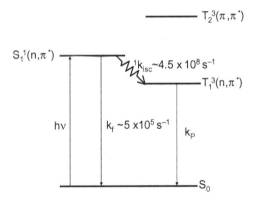

Figure 5.5 State diagram for acetone. Notice that intersystem crossing occurs between (n,π^*) states

If $^1k_{isc} \gg {}^1k_r$, any photochemical reactions from S_1 occur with very small quantum yield. If $^1k_{isc} \gg {}^1k_f$ then no fluorescence is observed. $^1k_{isc}$ is comparable with ^1k_f, so the T_1 state will always be formed and phosphorescence observed unless $^3k_r \gg k_p$.

The magnitude of $^1k_{isc}$ is governed by **El-Sayed's selection rules**; that is, the rate of intersystem crossing from the lowest singlet state to the triplet manifold is relatively large when the transition involves a change of orbital type. For example:

$$^1(n,\pi^*) \rightsquigarrow {}^3(\pi,\pi^*) \text{ is faster than } {}^1(n,\pi^*) \rightsquigarrow {}^3(n,\pi^*)$$
$$^1(\pi,\pi^*) \rightsquigarrow {}^3(n,\pi^*) \text{ is faster than } {}^1(\pi,\pi^*) \rightsquigarrow {}^3(\pi,\pi^*)$$

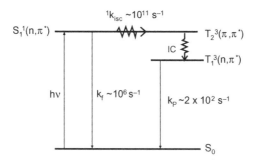

Figure 5.6 State diagram for benzophenone. Here intersystem crossing involves a change in orbital type

The slower processes occur with rate constants 10^{-2}–10^{-3} times those of the faster processes.

It is because of El-Sayed's selection rules that there is a great difference in luminescence behaviour between aliphatic and aromatic ketones. The principal reason for the difference is that $^1k_{isc}$ for aromatic ketones involves a change in orbital type whereas $^1k_{isc}$ for aliphatic ketones does not undergo a change in orbital type (Figures 5.5 and 5.6). Thus $^1k_{isc}$ is much greater for the aromatic ketones because of application of El-Sayed's rules.

6

Intermolecular Physical Processes of Excited States

AIMS AND OBJECTIVES

After you have completed your study of all the components of Chapter 6, you should be able to:

- Explain the deactivation of excited states by other molecules in terms of quenching processes, excimer/exciplex formation, energy transfer and electron transfer.
- Compare the effectiveness of different quenchers by means of the Stern–Volmer equation.
- Recognise situations in which the formation of excimers/exciplexes may affect the observed properties of an excited system.
- Understand that in photoinduced electron transfer, the excited state can be an electron donor or an electron acceptor, whereas in energy transfer the excited state is exclusively an energy donor.
- Understand that efficient energy transfer requires that the energy of an excited donor must be greater than that of the acceptor.
- Understand that the energetics of photoinduced electron transfer depends both on the redox potentials of the donor and acceptor and on the energy of the excited state.
- Understand that nonradiative singlet–singlet transfer can occur through either a dipole (Förster) mechanism, which can operate over long distance, or an exchange (Dexter) mechanism, which involves close approach.

Principles and Applications of Photochemistry   Brian Wardle
© 2009 John Wiley & Sons, Ltd

- Understand that both mechanisms for singlet–singlet energy transfer require an overlap between the florescence spectrum of the donor and the absorption spectrum of the acceptor. In addition, understand that a favourable orientation of donor and acceptor is necessary for the dipole mechanism.
- Understand that triplet–triplet energy transfer by the dipole mechanism is forbidden, but that the application of the Wigner spin rule shows that it can occur by the exchange mechanism.
- Understand the application of FRET to the dynamic processes in living cells, the measurement of distances in biological macromolecules and the recognition of specific nucleotide sequences in DNA samples.
- Understand the principles behind the use of photosensitised processes and relate these to the use of sensitised reactions in organic photochemistry and photodynamic therapy.
- Use simple molecular orbital diagrams to show the various processes of photoinduced energy transfer and photoinduced electron transfer.
- Apply the Marcus theory to photoinduced electron transfer and show how experimental evidence gives a relationship between the kinetics of the process and the thermodynamic driving force.

6.1   QUENCHING PROCESSES

The intramolecular processes responsible for radiative and radiationless deactivation of excited states we have considered so far have been unimolecular processes; that is, the processes involve only one molecule and hence follow first-order kinetics.

We now look at the intermolecular deactivation of an excited molecule by another molecule (of the same or different type), a process called **quenching**. Any substance that increases the rate of deactivation of an electronically-excited state is known as a **quencher** and is said to **quench** the excited state.

By measuring the reduction in fluorescence intensity in the presence of a quencher it is possible to determine the effect of quenching on the S_1 state.

Molecular oxygen is a very efficient quencher, such that in any quantitative work it is necessary to exclude oxygen either by bubbling oxygen-free nitrogen through the solution or by carrying out a number of freeze-pump-thaw cycles (cooling in liquid nitrogen, evacuating to remove any gas, sealing from the atmosphere and thawing to release dissolved gases).

Fluorescence quantum yields approach their maximum values when measured in very dilute solution from which all impurities are rigorously excluded. If the concentration of the fluorescent compound is increased, or if other substances are present in the solution, the fluorescence quantum yield will be reduced.

Consider the deactivation processes of the excited S_1 state in the presence of a quencher, Q:

$$S_1 \rightarrow S_0 + h\nu: \text{fluorescence}$$
$$S_1 \rightsquigarrow T_1: \text{intersystem crossing}$$
$$S_1 \rightsquigarrow T_1: \text{intersystem crossing}$$
$$S_1 + Q \rightsquigarrow S_0 + Q: \text{quenching.}$$

In the absence of the quencher, the unimolecular processes dominate:

$$^1J_{total} = (k_f + k_{isc} + k_{ic})[S_1] = {}^1k_{total}[S_1]$$

Quenching is a bimolecular process; that is, it involves the collision of both S_1 and Q molecules. Thus, in the presence of the quencher, where the rate constant is k_Q and the rate of deactivation by quenching is ^QJ:

$$^QJ = k_Q[S_1][Q]$$

Therefore, the overall rate of deactivation of S_1, $^QJ_{total}$, is given by the sum of the rates of the unimolecular and bimolecular processes:

$$^QJ_{total} = {}^1J_{total} + {}^QJ = {}^1k_{total}[S_1] + k_Q[S_1][Q]$$

Now, the rate of fluorescence emission is $J_f = k_f [S_1]$, and since the fluorescence quantum yield was shown in Section 3.5 to be equal to the ratio of the rate of fluorescence to the total rate of deactivation, the fluorescence quantum yields in the presence and absence of a quencher, $^Q\phi_f$ and ϕ_f respectively, are:

$$^Q\phi_f = J_f / {}^QJ_{total} = k_f[S_1] / ({}^1k_{total} + k_Q[S_1][Q]) = k_f / ({}^1k_{total} + k_Q[Q])$$

and:

$$\phi_f = k_f / {}^1k_{total}$$

The ratio of the two quantum yields leads to the **Stern–Volmer equation**:

$$\phi_f / {}^Q\phi_f = \left({}^1k_{total} + k_Q[Q]\right) / {}^1k_{total} = 1 + \left(k_Q[Q] / {}^1k_{total}\right)$$

It was shown in Section 3.5 that the excited-state lifetime of the S_1 state is given by:

$$^1\tau = 1 / {}^1k_{total}$$

Thus:

$$\phi_f / {}^Q\phi_f = 1 + k_Q{}^1\tau[Q]$$

or:

$$\boxed{\phi_f / {}^Q\phi_f = 1 + K_Q[Q]}$$

where K_Q is the **Stern–Volmer quenching constant**.

If the fluorescence quantum yield is measured in the absence and presence of known concentrations of quencher, a **Stern–Volmer plot** of $\phi_f/{}^Q\phi_f$ against [Q] will give a straight line of slope K_Q and intercept 1 (Figure 6.1).

Since fluorescence intensity and fluorescence lifetime are both proportional to the fluorescence quantum yield, plots of $I_f/{}^QI_f$ and $\tau_0/{}^Q\tau$ against [Q] are also linear with slope K_Q and intercept 1.

The Stern–Volmer method of monitoring the quenching of luminescence can also be applied to the quenching of phosphorescence emission.

6.2   EXCIMERS

When a dilute solution of the polynuclear aromatic hydrocarbon pyrene in toluene is irradiated with ultraviolet radiation, it emits a violet

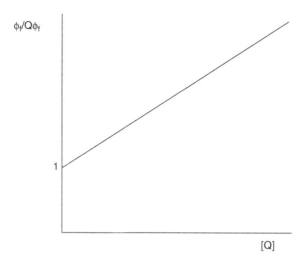

$\phi_f/Q\phi_f$

1

[Q]

Figure 6.1 A Stern–Volmer plot of fluorescence quenching

fluorescence. In a concentrated solution, the violet fluorescence is replaced by an intense sky-blue fluorescence. The fluorescence spectra of the two solutions are shown in Figure 6.2.

As the pyrene concentration increases, the following effects are observed:

- The emission due to pyrene at lower wavelength decreases in intensity.
- A new fluorescent emission appears at longer wavelength and increases in intensity.

The isoemissive point is indicative of the involvement of only two species in the observed fluorescence.

The explanation for this behaviour is that at the higher concentration of pyrene an excited-state dimer, or **excimer**, is formed through an interaction between the electronically-excited pyrene (M*) and the ground-state pyrene (M):

$$^1M^* + M \rightarrow {}^1[MM]^*$$
$$\downarrow \qquad\qquad \downarrow$$
$$M + h\nu \qquad M + M + h\nu'$$

The structure of the pyrene excimer is sandwich-shaped, with the distance between the planes of the two rings being of the order of 0.35 nm (Figure 6.3).

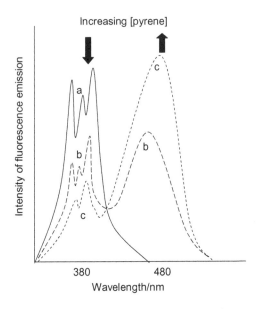

Figure 6.2 Fluorescence spectra of solutions of pyrene in toluene: (a) 10^{-6} mol dm^{-3}, (b) 10^{-4} mol dm^{-3}, (c) 10^{-3} mol dm^{-3}
Adapted from T. Forster, 'Excimers', *Angewandte Chemie International Edition,* Volume 8, no. 5 (1969), Verlag-Chemie

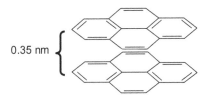

Figure 6.3   'Sandwich' structure of the pyrene excimer

The formation of such excimers, which only exist in the excited state, is commonplace among polynuclear aromatic hydrocarbons, the simple potential energy diagram for which is shown in Figure 6.4.

- The lower of the two potential energy curves in the figure, corresponding to the ground-state molecules, is characteristic of a dissociative system in which no bound dimer is formed.
- Bringing the M* and M molecules together, however, results in a potential energy minimum at an internuclear distance, shown as r_{eq}. This minimum corresponds to the formation of the bound excimer.

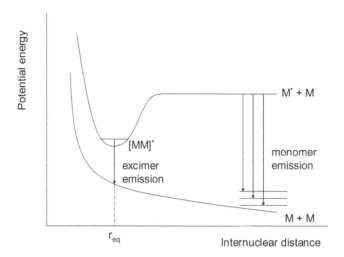

Figure 6.4 Potential energy diagram for a typical excimer, showing the ground state and first excited singlet state

- The energy change associated with the excimer emission is smaller than that for the monomer emission and so the excimer emission will occur at longer wavelengths than the monomer emission.
- Vibrational fine structure is absent from the excimer emission because the Franck–Condon transition is to the unstable dissociative state where the molecule dissociates before it is able to undergo a vibrational transition. In the case of the monomer emission, all electronic transitions are from the $v = 0$ vibrational level of M^* to the quantised vibrational levels of M, resulting in the appearance of vibrational fine structure.

6.2.1 Excimer Emission in Ca^{2+} Sensing

Fluorescent molecule (1) (Figure 6.5) acts as a host for Ca^{2+} ions. On addition of Ca^{2+} ions, the flexible polyether chain folds, leading to stacking of the anthracene fluorophores, and monomer emission is replaced by excimer emission.

6.3 EXCIPLEXES

Phenomena analogous to those described for excimers are observed in solutions containing two compounds. For example, if N,N-diethylaniline

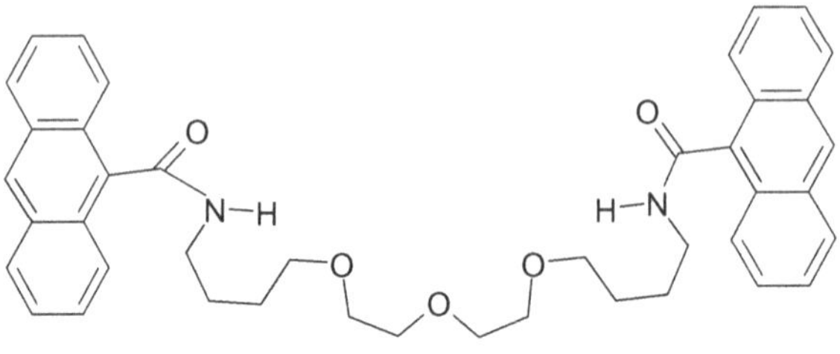

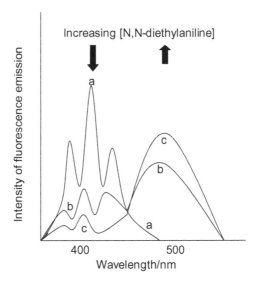

Figure 6.5   Fluorescent molecule (1)

Figure 6.6   Fluorescence spectra of anthracene in toluene in the presence of N,N-diethylaniline of varying concentration

is added to a solution of anthracene in a nonpolar solvent such as toluene, the anthracene fluorescence is quenched and replaced by a structureless emission at longer wavelength (Figure 6.6). An excited complex, or **exciplex**, is formed by reaction between the electronically-excited anthracene molecule (M^*) and the ground-state N,N-diethylaniline molecule, which effectively acts as a quencher (Q).

$$M^* + Q \rightarrow [MQ]^*$$
$$\downarrow \qquad\qquad \downarrow$$
$$M + h\nu \qquad M + Q + h\nu'$$

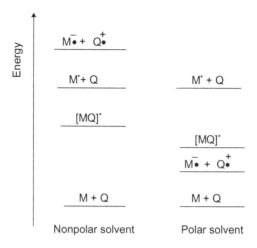

Figure 6.7 Energy diagram for exciplex formation in solvents of differing polarities
Adapted from A. Gilbert and N. Baggott, *Essentials of Molecular Photochemistry* (London: Blackwell, 1991)

The chemical association of the exciplex results from an attraction between the excited-state molecule and the ground-state molecule, brought about by a transfer of electronic charge between the molecules. Thus exciplexes are polar species, whereas excimers are nonpolar. Evidence for the charge-transfer nature of exciplexes in nonpolar solvents is provided by the strong linear correlation between the energy of the photons involved in exciplex emission and the redox potentials of the components.

The exciplex emission is also affected by solvent polarity, where an increase in the solvent polarity results in a lowering of the energy level of the exciplex, at the same time allowing stabilisation of charged species formed by electron transfer (Figure 6.7). Thus, in polar solvents the exciplex emission is shifted to even higher wavelength and accompanied by a decrease in the intensity of the emission, due to competition between exciplex formation and electron transfer.

6.3.1   Exciplex Fluorescence Imaging

The development of efficient combustion systems with reduced emissions can be brought about by having a clear understanding of the fuel : air mixing ratios in the combustion chamber. Concentration distributions of the liquid and vapour phases of the fuel can be imaged by doping the fuel with a chemical that has similar properties to the fuel. Such chemicals are designed to fluoresce at different wavelengths in each fuel phase when irradiated with ultraviolet light.

Exciplex formation occurs only in the liquid phase, where a green emission occurs on irradiation, while in the vapour phase a blue emission is observed because of the monomer. Using careful calibration techniques it is possible to perform accurate concentration measurements of the fuel/air mixtures.

6.4   INTERMOLECULAR ELECTRONIC ENERGY TRANSFER

From the photochemical point of view, intermolecular electronic energy transfer may be regarded as being the transfer of energy from the excited state of one molecular entity (the donor, D^*) to another (the acceptor, A). The excited state of the donor is deactivated to a lower-lying electronic energy state by transferring energy to the acceptor, which is thereby raised to a higher electronic energy state:

$$D^* + A \rightarrow D + A^*$$

The nomenclature relating to electronic energy transfer is such that it is named according to the multiplicity of D^* and A^* as D^*–A^* energy transfer, common examples being:

singlet–singlet energy transfer: $^1D^* + {}^1A \rightarrow {}^1D + {}^1A^*$
triplet–triplet energy transfer: $^3D^* + {}^1A \rightarrow {}^1D + {}^3A^*$.

Since the excited-state donor molecules are initially produced by photoexcitation and the energy is transferred to A, energy transfer is also referred to as the **photosensitisation** of A or the quenching of D^*.

The principal mechanisms by which electronic energy transfer occurs are shown in Figure 6.8.

Nonradiative energy transfer has a major role in the process of photosynthesis. Light is absorbed by large numbers of chlorophyll molecules in light-harvesting antennae and energy is transferred in a stepwise manner to photosynthetic reaction centres, at which photochemical reactions occur. This fundamental energy-transfer process will be considered in more detail in Chapter 12.

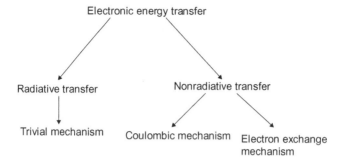

Figure 6.8   The principal mechanisms of electronic energy transfer. All three mechanisms require overlap between the fluorescence spectrum of the donor and the absorbance spectrum of the acceptor

Energy transfer has also found extensive application in understanding the dynamic processes that occur in biological cells and in determining the size and shape of biological macromolecules. Research at the forefront of cancer diagnosis using **molecular beacons** also relies on an understanding of energy-transfer processes.

6.5   THE TRIVIAL OR RADIATIVE MECHANISM OF ENERGY TRANSFER

This process involves the emission of a quantum of light by an excited donor, followed by the absorption of the emitted photon by the acceptor.

$$D^* \rightarrow D + h\nu$$
$$h\nu + A \rightarrow A^*$$

Thus, this mechanism requires that A must be capable of absorbing the photon emitted by D^*; that is, the acceptor absorption spectrum must overlap with the donor emission spectrum. Radiative energy transfer can operate over very large distances because a photon can travel a long way and A simply intercepts the photon emitted by D^*.

Since the photon emitted by D^* is absorbed by A, the same rules will apply to radiative energy transfer as to the intensity of absorption. Because singlet–triplet transitions are spin-forbidden and singlet–triplet absorption coefficients are usually extremely small, it is not possible to build up a triplet state population by radiative energy transfer. For this reason, radiative energy transfer involving the formation of a triplet excited state of $^3A^*$ does not take place.

6.6   LONG-RANGE DIPOLE–DIPOLE (COULOMBIC) ENERGY TRANSFER

The Coulombic mechanism is a relatively long-range process in as much as energy transfer can be significant even at distances of the order of 10 nm.

Coulombic energy transfer is a consequence of mutual electrostatic repulsion between the electrons of the donor and acceptor molecules. As D* relaxes to D, the transition dipole thus created interacts by Coulombic (electrostatic) repulsion with the transition dipole created by the simultaneous electronic excitation of A to A* (Figure 6.9).

In order for rapid and efficient energy transfer to occur by the Coulombic mechanism:

- there should be good spectral overlap between the fluorescence spectrum of the donor and the absorption spectrum of the acceptor
- the transition corresponding to the donor emission should be allowed (large ϕ_f)
- the relevant overlapping absorption band of the acceptor should be strong (large ε_A)
- a favourable mutual orientation of the transition dipoles should exist.

The Coulombic mechanism can only occur where spin multiplicity is conserved since it is only in such transitions that large transition dipoles

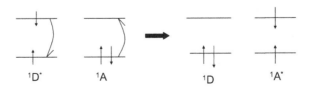

Figure 6.9   Electron movements occurring in long-range Coulombic energy transfer. Note that the electrons initially on D* remain on D and electrons initially on A remain on A*. This energy transfer does not require physical contact between the donor and acceptor

are produced. Singlet–singlet energy transfer occurs by this mechanism as the donor (excited singlet to singlet) and acceptor (singlet to excited singlet) undergo no change in multiplicity, resulting in the creation of significant transition dipoles:

$$^1D^* + {}^1A \rightarrow {}^1D + {}^1A^*$$

However, triplet–triplet energy transfer cannot occur by this mechanism as this would require both donor and acceptor to undergo a change in multiplicity:

$$^3D^* + {}^1A \rightarrow {}^1D + {}^3A^*$$

Coulombic energy transfer is sometimes called **resonance energy transfer** because the energies of the coupled transitions are identical, or in other words, in resonance (Figure 6.10).

A detailed theory of energy transfer by the Coulombic mechanism was developed by Förster, so the process is often referred to as **Förster resonance energy transfer** (**FRET**). According to the Förster theory, the probability of Coulombic energy transfer falls off inversely with the sixth power of the distance between the donor and the acceptor. For

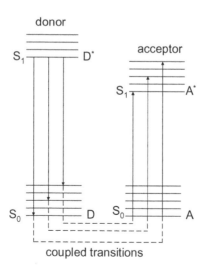

Figure 6.10 Energy-level diagram showing the coupling of donor and acceptor transitions of equal energy required for long-range nonradiative transfer

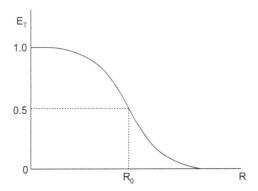

Figure 6.11   The dependence of the efficiency of energy transfer, E_T, on the donor-acceptor distance, R, according to the Förster theory

donor–acceptor pairs, the efficiency of resonance energy transfer, E_T, increases with decreasing distance, R, according to:

$$E_T = R_0^6 / (R_0^6 + R^6)$$

where R_0 is the **critical transfer distance**, which is characteristic for a given donor–acceptor pair. R_0 is the donor–acceptor distance at which energy transfer from D* to A and internal deactivation are equally probable; that is, R_0 is the donor–acceptor distance at which the efficiency of energy transfer is 50% (Figure 6.11).

The efficiency of energy transfer can also be determined using:

- The relative fluorescence intensity of the donor in the absence (F_D) and the presence (F_{DA}) of the acceptor:

$$E_T = 1 - (F_{DA} / F_D)$$

- The relative fluorescence quantum yield of the donor in the absence (ϕ_D) and the presence (ϕ_{DA}) of the acceptor:

$$E_T = 1 - (\phi_{DA} / \phi_D)$$

- The relative fluorescence lifetime of the donor in the absence (τ_D) and the presence (τ_{DA}) of the acceptor:

$$E_T = 1 - (\tau_{DA} / \tau_D)$$

When it is found experimentally that the rate constant for energy transfer is both insensitive to solvent viscosity and significantly greater than the rate constant for diffusion then the Coulombic mechanism is confirmed. The process is equivalent to the energy being transferred across the space between donor and acceptor, rather like a transmitter–antenna system. As D* and A are brought together in solution, they experience long-range Coulombic interactions due to their respective electron charge clouds.

6.6.1   Dynamic Processes within Living Cells

Energy-transfer measurements are often used to estimate the distances between sites on biological macromolecules such as proteins, and the effects of conformational changes on these distances. In this type of experiment, the efficiency of energy transfer is used to calculate the distance between donor and acceptor fluorophores in order to obtain structural information about the macromolecule.

Green fluorescent protein (GFP) was originally isolated from a jellyfish that fluoresces green when exposed to blue light. Incorporation of the genetic information relating to GFP has enabled the development of fluorescent biomarkers in living cells. Several spectrally-distinct mutation variants of the protein have been developed, such as blue fluorescent protein (BFP), which emits blue light. The absorption and fluorescence spectra for GFP and BFP are such that they allow FRET measurements to be used to determine donor–acceptor distances within biomolecules.

Consider Förster resonance energy transfer between a donor and an acceptor attached to opposite ends of the same macromolecular protein. In the normal conformation the donor and acceptor are separated by a distance too great for FRET to occur, but after undergoing a conformational change the donor and acceptor are brought closer together, enabling FRET to occur. Figure 6.12 shows a protein labelled with BFP (the donor) and GFP (the acceptor). The acceptor emission maximum (510 nm) will be observed when the complex is excited at the maximum absorbance wavelength (380 nm) of the donor, provided that the distance between the BFP and GFP will allow FRET to occur.

Advances in pulsed lasers, microscopy and computer imaging, and the development of labelling techniques in which the donor and acceptor fluorophores become part of the biomolecules themselves have enabled the visualisation of dynamic protein interactions within living cells.

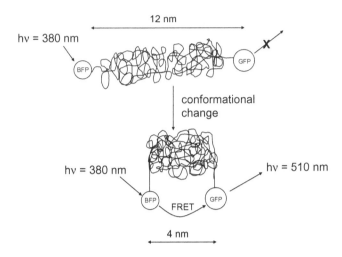

Figure 6.12   Labelling a protein molecule with a donor and acceptor group to show the dependence of FRET on distance

6.6.2   Molecular Ruler

FRET manifests itself through the quenching of donor fluorescence and a reduction of the fluorescence lifetime, accompanied by an increase in acceptor fluorescence emission. The efficiency of the energy-transfer process varies in proportion to the inverse sixth power of the distance separating the donor and acceptor molecules. Consequently, FRET measurements can be utilised as an effective **molecular ruler** for determining distances between molecules labelled with an appropriate donor and acceptor fluorophore, provided they are within 10 nm of each other.

If the donor and acceptor group are separated by spacer groups giving varying donor–acceptor distances, R, then the efficiency of energy transfer, E_T, can be determined by measuring the fluorescence intensity in the absence and presence of the acceptor group. Plotting the data as shown in Figure 6.9 allows calculation of the critical transfer distance. To confirm that the energy transfer process occurs by the Coulombic mechanism, a plot of $\ln(E_T^{-1} - 1)$ against $\ln R$ will be linear with slope ~6 if the results are in agreement with the R^{-6} dependence predicted by Förster theory.

6.6.3   Molecular Beacons

In molecular biology, a set of two hydrogen-bonded nucleotides on opposite complementary nucleic acid strands is called a base pair. In the classical Watson–Crick base pairing in DNA, adenine (A) always forms a base pair with thymine (T) and guanine (G) always forms a base pair with cytosine (C). In RNA, thymine is replaced by uracil (U).

Cancer cells develop from alterations in genes (composed of nucleic acids), which confer growth advantage and the ability of the cancer to spread to different parts of the body. A novel way of achieving early detection of cancer is to detect nucleotide sequences of cancer-causing genes in living cells.

Molecular beacons, developed in the mid-1990s, can be delivered into cells with high efficiency, where they are able to detect particular sequences of nucleotides that are indicative of certain types of cancers, thus aiding the early diagnosis of the disease.

Molecular beacons are single-stranded hairpin-shaped nucleotide probes. In the presence of the target nucleotide sequence the molecular beacon unfolds, binds and fluoresces (Figure 6.13).

The molecular beacon essentially consists of four parts:

- Loop: this is the nucleotide region of the molecular beacon, which is complementary to the target nucleotide sequence.
- Stem: the beacon stem sequence lies on both the ends of the loop. It is a few complementary base pairs long.

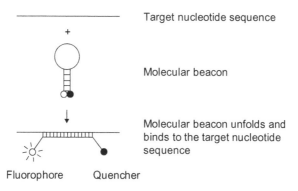

Target nucleotide sequence

Molecular beacon

Molecular beacon unfolds and binds to the target nucleotide sequence

Fluorophore   Quencher

Figure 6.13   Mode of action of a molecular beacon. Fluorescence is only observed when the beacon and the target nucleotide bind together

- Fluorophore: toward one end of the stem a fluorescent group is attached.
- Quencher: the other end of the stem is attached to a group capable of quenching the fluorophore. When the beacon is in the unbound state, the close proximity of the quencher and fluorophore prevents the emission of light from the fluorophore.

Even though it is possible to construct molecular beacons that fluoresce in a variety of different colours, the number of different molecular beacons that can be used in the same solution to simultaneously detect different targets is limited. Monochromatic light sources excite different fluorophores to different extents. In order to overcome this limitation, a series of different molecular beacons is formed in which each probe emits visible fluorescence of specific wavelength, yet at the same time each undergoes efficient excitation by the same monochromatic light source. In such probes, the absorption of energy from light and the emission of that energy as fluorescence are performed by two separate fluorophores. Because the emission spectrum is shifted from the characteristic emission range of the fluorophore that absorbs the incident light to the characteristic emission range of a second fluorophore, these probes are called wavelength-shifting molecular beacons (Figure 6.14).

A wavelength-shifting molecular beacon contains three labels:

- Quencher moiety (black circle), on one end of the stem.

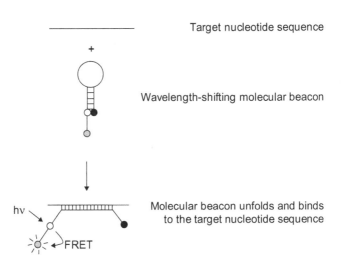

Figure 6.14 Principle of operation of a wavelength-shifting molecular beacon

- Harvester fluorophore (white circle), on the opposite end of the stem to the quencher in the hairpin conformation.
- Emitter fluorophore (grey circle), attached to the same end of the hairpin conformation as the harvester fluorophore.

The harvester fluorophore is chosen so that it efficiently absorbs energy from the monochromatic light source. In the absence of targets, these probes emit no light because the energy absorbed by the harvester fluorophore is rapidly transferred to the quencher and is lost as heat. In the presence of targets, molecular beacons undergo a conformational reorganisation caused by the rigidity of the probe–target hybrid, which forcibly separates the arms. In the target-bound conformation, the energy absorbed by the harvester fluorophore is transferred by FRET to the emitter fluorophore, which then releases the energy as fluorescent light of its own characteristic colour.

A simple diagnostic test has been devised for prostate cancer, using a specific molecular beacon mixed with the target DNA on a microscope slide. The DNA is treated to separate the strands and, provided there is the correct correspondence between the bases, *in situ* combination occurs between the bases on the molecular beacon and those on the strands. Thus, if fluorescence is observed, the DNA sample must contain the base sequence indicative of prostate cancer.

6.7   SHORT-RANGE ELECTRON-EXCHANGE ENERGY TRANSFER

Consider triplet–triplet energy transfer:

$$^3D^* + {}^1A \rightarrow {}^1D + {}^3A^*$$

The Coulombic mechanism would require that both $^3D^* \rightarrow {}^1D$ and $^1A \rightarrow {}^3A^*$ were allowed transitions, which clearly they are not as both are spin-forbidden processes. Thus, triplet–triplet energy transfer by the long-range Coulombic mechanism is forbidden.

As the donor and acceptor molecules approach each other closely so that their regions of electron density overlap, electrons can be exchanged between the two molecules. This mechanism is therefore called the **exchange mechanism**. The electron-exchange mechanism requires a close approach (1–1.5 nm), though not necessarily actual contact,

Table 6.1 Some energy-transfer processes by the electron-exchange mechanism allowed according to the Wigner spin conservation rule

Energy Transfer	Spin States Before and After Energy Transfer	
singlet–singlet	spin states	$1D^* + 1A \rightarrow 1D + 1A^*$ ↑↓↑↓ ↑↓↑↓
		0 0
triplet–triplet	spin states	$3D^* + 1A \rightarrow 1D + 3A^*$ ↑↑↑↓ ↑↓↑↑
		1 1
singlet quenching by oxygen	spin states	$1D^* + 3O2 \rightarrow 3D^* + 1O2$ ↑↓↑↑ ↑↑↑↓
		1 1
triplet quenching by oxygen	spin states	$3D^* + 3O2 \rightarrow 1D^* + 1O2$ ↑↑↓↓ ↑↓↑↓
		0 0

between the donor and the acceptor in order to facilitate physical transfer of electrons between the two.

Energy transfer by the exchange mechanism will occur provided the spin states before and after overlap obey the **Wigner spin conservation rule**; that is, provided the overall spin states before and after overlap have common components (Table 6.1).

According to the **Dexter theory of energy transfer**, the distance dependence of energy transfer by the exchange mechanism falls off rapidly and is given by:

$$k_{ET(exchange)} \propto \exp(-2R_{DA})$$

From the molecular orbital point of view, unlike the Coulombic mechanism, electronic energy transfer by the exchange mechanism requires transfer of electrons between the donor and acceptor molecules. Figure 6.15 shows the molecular orbital energy diagram of an electron exchange between D^* and A. The two electron-transfer processes occur simultaneously so that both the donor and the acceptor remain as uncharged species throughout the exchange process.

6.7.1   Triplet–Triplet Energy Transfer and Photosensitisation

Excitation of benzophenone in solid solution at 77 K with light of wavelength 366 nm produces phosphorescence. As naphthalene is added, the

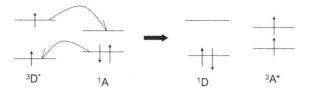

Figure 6.15 Electron movements occurring in short-range triplet–triplet energy transfer by the exchange mechanism. Note that an electron initially on D* moves to A and an electron initially on A moves to D*

benzophenone phosphorescence is quenched and replaced with phosphorescence emission from the naphthalene, even though naphthalene does not absorb photons of wavelength 366 nm. These observations are accounted for by the following sequence of events:

- Light is absorbed by the benzophenone:

$$^1(C_6H_5)_2\,CO + h\nu \rightarrow\ ^1(C_6H_5)_2CO^*$$

- Intersystem crossing to the benzophenone triplet then occurs:

$$^1(C_6H_5)_2CO^* \rightsquigarrow\ ^3(C_6H_5)_2CO^*$$

- Triplet–triplet energy transfer to naphthalene occurs as the energy of:

$$^3(C_6H_5)_2\,CO^* > \ ^3C_{10}H_8^*$$
$$^3(C_6H_5)_2\,CO^* + \ ^3C_{10}H_8 \rightarrow\ ^1(C_6H_5)_2\,CO + \ ^3C_{10}H_8^*$$

- Phosphorescence emission occurs from the triplet naphthalene:

$$^3C_{10}H_8^* \rightarrow\ ^1C_{10}H_8 + h\nu$$

The process whereby a photophysical or photochemical change occurs in one molecule as the result of light absorption by another is known as **photosensitisation**. The molecule which initially absorbs the light in order to bring the change about is called a **photosensitiser** (Figure 6.16). Triplet–triplet energy transfer allows the efficient indirect production of triplet molecules, which are incapable of being produced directly due to inefficient intersystem crossing.

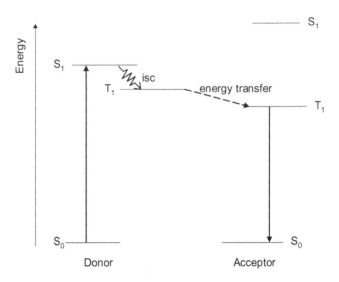

Figure 6.16   The process of triplet–triplet energy transfer between a donor (photosensitiser) and an acceptor. The triplet photosensitiser transfers its energy to the acceptor

To prevent direct excitation of the acceptor molecule, the lowest excited singlet state of the donor should be lower in energy than the lowest excited singlet state of the acceptor. For triplet–triplet energy transfer to occur, the lowest triplet state of the donor must be higher in energy than the lowest triplet state of the acceptor. Because ketones have a small singlet–triplet energy gap they appear to be well suited as photosensitisers. Their high triplet energies and high triplet quantum yields make them excellent triplet photosensitisers (see Section 8.3).

6.7.2   Singlet Oxygen and Photodynamic Therapy for Cancer Treatment

Whereas most ground-state molecular species encountered are singlet states (^1R), a simple molecular orbital treatment shows the ground-state oxygen molecule to be a triplet ($^3O_2$). The lowest excited state of O_2 is a low-energy singlet state ($^1O_2^*$), which cannot be obtained by direct irradiation of the triplet ground state because it is a spin-forbidden process. However, a photosensitisation process can be used, although it is in the reverse direction of what is usually seen in sensitised reactions, involving triplet-ground-state-to-singlet-excited-state sensitisation.

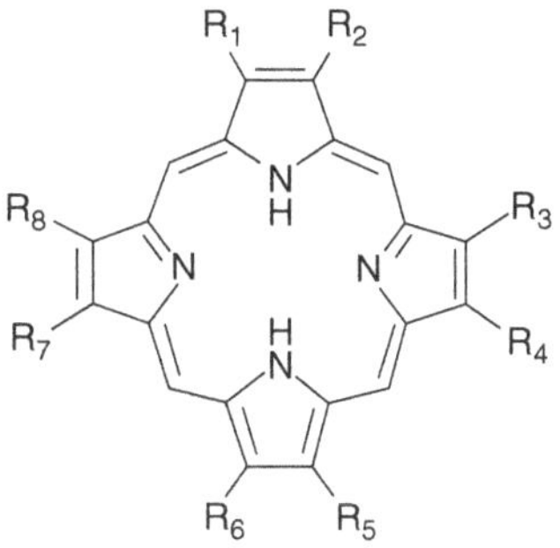

Figure 6.17 Generalised porphyrin structure. The R groups represent different side groups attached to the porphyrin ring

$$^1D + h\nu \rightarrow {}^1D^* \rightsquigarrow {}^3D^*$$
$$^3D^* + {}^3O_2 \rightarrow {}^1D + {}^1O_2^*$$

Any molecule with an excited triplet energy greater than the energy of $^1O_2^*$ is capable of bringing about the sensitisation reaction, but molecules with low excited-triplet-state energies are preferred as they avoid the possibility of unwanted side reactions. Molecules which absorb in the visible region, such as rose bengal, methylene blue and porphyrins, can act as efficient sensitisers of singlet oxygen.

Photodynamic therapy for the treatment of tumours involves the selective uptake and retention of a highly-coloured porphyrin sensitiser (Figure 6.17) in the tumour. Irradiation by a laser with a wavelength corresponding to the absorption maximum of the porphyrin (D) causes excitation of the porphyrin to the excited singlet state.

$$^1D + h\nu \rightarrow {}^1D^*$$

Intersystem crossing results in the formation of the excited triplet state of the porphyrin:

$$^1D^* \rightsquigarrow {}^3D^*$$

The porphyrin excited triplet can then undergo:

- Photochemical reaction with organic molecules within the tumour cells, where hydrogen abstraction occurs. This initiates a number of radical reactions, resulting in destruction of the tumour.

- Energy transfer to $^3O_2$, which is found in most cells, resulting in production of singlet oxygen, $^1O_2^*$, provided the energy of $^3D^*$ is greater than that of $^1O_2^*$.

$$^3D^* + {}^3O_2 \rightarrow {}^1D + {}^1O_2^*$$

Singlet oxygen, $^1O_2$, is highly toxic and oxidises organic substrates found within the tumour cells, destroying the tumour in the process.

6.8   PHOTOINDUCED ELECTRON TRANSFER (PET)

Almost all life on Earth relies on the photosynthesis processes of bacteria or higher plants. In the photosynthetic reaction centre, sunlight triggers a chain of electron-transfer reactions, which results in effective charge separation and ultimately leads to a proton gradient across a biological membrane. As a result, free energy originating from the light quanta is stored, and this can be utilised for the generation of energy-rich adenosine triphosphate (ATP). These PET processes relating to photosynthesis will be considered in more detail in Chapter 12.

At this point we need to be conversant with how the redox properties of molecules are affected by the absorption of light. Absorption of a photon by a molecule promotes an electron to a higher energy level and the molecule becomes a better electron donor (reducing agent) in its excited state than in its ground state. As electronic excitation also creates an electron vacancy in the highest occupied molecular orbital, the molecule is also a better electron acceptor (oxidising agent) in its excited state.

PET corresponds to the primary photochemical process of the excited-state species, $R^* \rightarrow I$, where R^* can be an electron donor or electron acceptor when reacting with another molecule, M.

- R^* acts as an electron donor and is therefore oxidised (oxidative electron transfer; Figure 6.18):

$$R^* + M \rightarrow R^{\cdot+} + M^{\cdot-}$$

- R^* acts as an electron acceptor and is therefore reduced (reductive electron transfer; Figure 6.19):

$$R^* + M \rightarrow R^{\cdot-} + M^{\cdot+}$$

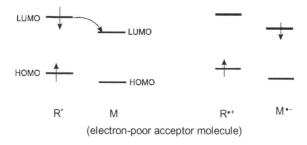

Figure 6.18   Molecular orbital representation of oxidative electron transfer

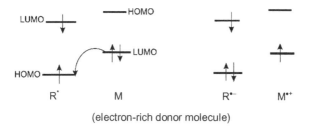

Figure 6.19   Molecular orbital representation of reductive electron transfer

In PET, an electron is transferred between an excited-state species and a ground-state species, and Figures 6.18 and 6.19 show that electron transfer processes occur by electron-exchange interactions and so require orbital overlap.

6.8.1   Fluorescence Switching by PET

The molecules used as fluorescent probes for cations, including H^+, are nonluminescent, but become fluorescent when bonded to the cation.

The probe molecules used consist of a fluorescent chromophore (fluorophore) bonded to a cation receptor group containing an electron-donating group. PET occurs from the receptor to the fluorophore, resulting in quenching of the latter (Figure 6.20). Binding of the receptor group to the cation restores the fluorescence properties of the fluorophore as electron transfer is inhibited. Thus the PET sensor acts as a molecular switch.

Figure 6.21 illustrates the principle of the process in terms of the molecular orbitals involved. Excitation of the fluorophore causes an

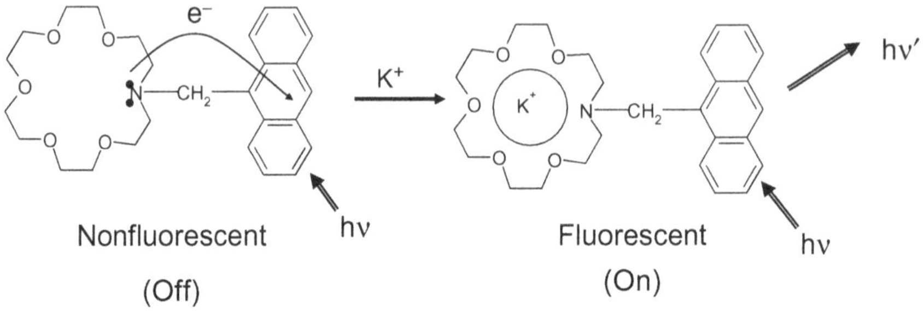

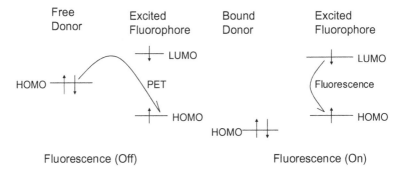

Figure 6.20   Action of a fluorescent PET potassium cation sensor as a molecular switch using a macrocyclic electron donor and anthracene fluorophore

Figure 6.21   Principle of the PET cation sensor

electron to be promoted from the HOMO to the LUMO. This enables PET to occur from the HOMO of the receptor to that of the fluorophore, resulting in fluorescence quenching of the latter. On binding with the K^+ cation, the redox potential of the electron-donating receptor is raised so that its HOMO becomes lower in energy than that of the fluorophore. Thus, as PET is no longer possible, fluorescence quenching does not occur.

6.8.2   The Marcus Theory of Electron Transfer

The basic assumption of the Marcus theory of electron transfer is that only a weak interaction between the reactants is needed in order for a simple electron-transfer process to occur. Marcus theory considers

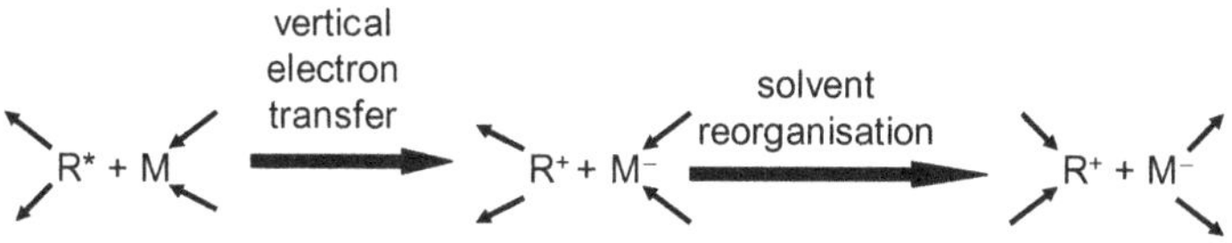

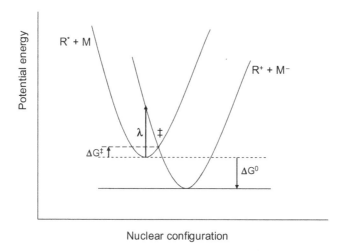

Figure 6.22 Rearrangement of polar solvent dipoles (arrows) during the electron transfer process $R^* + M \rightarrow R^+ + M^-$. The initial stage is a vertical (Franck–Condon) electron transfer, and this is followed by reorganisation of the solvent dipoles

Figure 6.23 Potential energy description of an electron-transfer reaction. The parabolic curves intersect at the transition state ($\ddagger$)

electron transfer in terms of reaction rate theory, potential energy surfaces and reorganisation of the system during electron transfer.

For a polar solvent, the solvent dipoles will be arranged around the various species taking part in the electron-transfer process. In the electron-transfer process $R^* + M \rightarrow R^+ + M^-$, solvent reorganisation may be necessary in order to accommodate and stabilise the charged species (Figure 6.22) and this process will require energy to be put into the system.

The potential energy surfaces on which the electron-transfer process occurs can be represented by simple two-dimensional intersecting parabolic curves (Figure 6.23). These quantitatively relate the rate of electron transfer to the reorganisation energy (λ) and the free-energy changes for the electron-transfer process (ΔG^0) and activation ($\Delta G^{\ddagger}$).

$\Delta G^{\ddagger}$ is the free energy of activation; that is, the free energy needed to reach the transition state, $\ddagger$. λ is the reorganisation energy, which corresponds to the total reorganisation of the donor, acceptor and solvent molecules during electron transfer. The free-energy change of the electron-transfer process, ΔG^{0}, is a measure of the driving force of the overall process.

From the geometry of parabolas:

$$\Delta G^{\ddagger} = \left(\Delta G^{0} \pm \lambda\right)^{2}\big/4\lambda$$

According to the collision theory, the rate constant (k_{ET}) of the electron-transfer process is given by:

$$k_{ET} = A\exp(-\Delta G^{\ddagger}/RT) = A\exp\left[-\left(\Delta G^{0} \pm \lambda\right)^{2}\big/4\lambda\right]\big/RT$$

giving:

$$\ln k_{ET} = \ln A + \left[-\left(\Delta G^{0} \pm \lambda\right)^{2}\big/4\lambda\right]\big/RT$$

This equation, based on the Marcus model, therefore gives us a relationship between the kinetics (k_{ET}) and the thermodynamic driving force (ΔG^{0}) of the electron-transfer process. Analysis of the equation predicts that one of three distinct kinetic regions will exist, as shown in Figure 6.24, depending on the driving force of the process.

- In the **normal region**, thermodynamic driving forces are small. The electron-transfer process is thermally activated, with its rate increasing as the driving force increases.
- In the **activationless region** (so called because $\Delta G^{\ddagger} = 0$), a change in the thermodynamic driving force has a negligible effect on the rate of electron transfer.
- The **inverted region** is so-called because the rate of the electron-transfer process decreases with increasing thermodynamic driving force.

6.8.3   Experimental Evidence Relating to the Marcus Equation

Research on photochemical reaction centres has been important for our understanding of the relationship between the structures and the rates

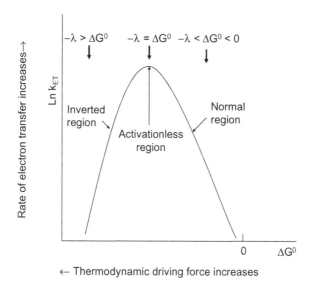

Figure 6.24 Marcus theory predictions of the dependence of the electron-transfer rate on the thermodynamic driving force

of electron transfer in proteins. The structures of such proteins can be determined by crystallography, and the rates of electron transfer can be accurately measured by picosecond transient spectroscopy (see Chapter 10). Rate constants for several different reactions of the photochemical cycle can be measured, and by using a variety of molecular modification procedures a range of values for ΔG° for the reactions can be obtained. These make it possible to show that the reactions behaved as expected from Marcus theory, and to derive a general rule known as **Dutton's ruler**, relating rate of reaction to the distance between the redox centres.

$$\log_{10} k_{ET} = 13 - 0.6(R - 3.6) - 3.1(\Delta G^\circ + \lambda)^2 / \lambda$$

where R is the edge-to-edge distance in angstroms.

The maximum rate of electron transfer occurs when $\lambda = -\Delta G^0$ and:

$$\log_{10} k_{ET(Max)} = 13 - 0.6(R - 3.6)$$

ΔG° for the single electron-transfer reaction can be obtained from the redox potentials of the systems involved in the process and is given by:

$$\Delta G^\circ = \Delta E^\circ_{(reductant-oxidant)}$$

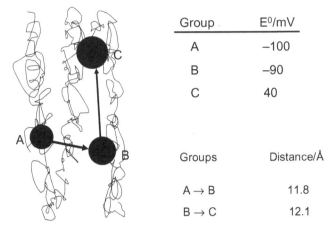

Group	E^0/mV
A	−100
B	−90
C	40

Groups	Distance/Å
A → B	11.8
B → C	12.1

Figure 6.25   Sequential electron transfer processes between protein-bound groups

For example, in the electron-transfer system shown in Figure 6.25 the protein-bound groups between which electron transfer occurs are designated A, B and C and the direction of electron transfer is shown as arrows.

A → B: The electron is lost by A and gained by B and so A is the reductant and B the oxidant. Therefore, $\Delta G^\circ = -100 + 90 = -10\,mV$

B → C: The electron is lost by B and gained by C and so B is the reductant and C the oxidant. Therefore, $\Delta G^\circ = -90 - 40 = -130\,mV$.

Using a typical λ value for intramolecular electron transfer of $750\,mV$, it is possible to use the Dutton's ruler expression to determine the rates of the two electron-transfer processes:

$$\log k_{ET(A\to B)} = 13 - 0.6(11.8 - 3.6) - 3.1(-0.01 + 0.75)^2 / 0.75 = 5.82$$

Thus:

$$k_{ET(A\to B)} = 6.6 \times 10^5 \text{ s}^{-1}$$

$$\log k_{ET(B\to C)} = 13 - 0.6(12.1 - 3.6) - 3.1(-0.13 + 0.75)^2 / 0.75 = 6.31$$

Thus:

$$k_{ET(A\to B)} = 2.0 \times 10^6 \text{ s}^{-1}$$

6.8.4   Evidence for the Inverted Region

A **dyad** is a supramolecular structure consisting of two distinct linked components such as donor and acceptor groups (see Chapter 12).

Fullerenes such as C_{60} are excellent electron acceptors. In a fullerene-porphyrin-based dyad, the photoexcited state of the C_{60} accepts an electron from the linked zinc porphyrin group to give a charge-separated state.

Using the dyad shown in Figure 6.26, it has been possible to investigate the driving-force dependence of the rate constants for the electron-transfer processes.

The PET process in the dyad is located in the normal region of the Marcus curve, while the back-electron transfer from C_{60}^{-} to $ZnPor^{+}$ is in the Marcus inverted region (Figure 6.27).

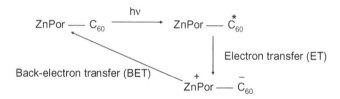

Figure 6.26   Dyad

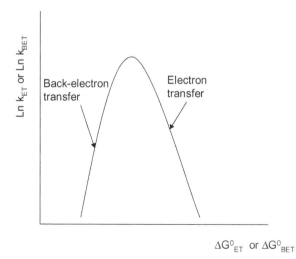

$$\Delta G^{0}_{ET} \text{ or } \Delta G^{0}_{BET}$$

Figure 6.27   Electron transfer processes in a zinc porphyrin-C_{60} dyad

7

Some Aspects of the Chemical Properties of Excited States

AIMS AND OBJECTIVES

After you have completed your study of all the components of Chapter 7, you should be able to:

- Use potential energy curves to explain the various types of behaviour when photons interact with diatomic molecules.
- Understand the differences between the reaction profiles of a photochemical reaction and a thermal reaction.
- Explain the difference between a concerted reaction and a stepwise reaction.
- Relate the principal differences between photochemical reactions and thermal reactions and explain these differences in terms of excited-state and ground-state species.
- Understand how photolysis produces radicals by bond cleavage and account for the importance of radical species in photochemical chain reactions, stratospheric ozone chemistry and the photochemistry of the polluted troposphere.
- Describe the principal unimolecular and bimolecular reactions of free radicals and explain the usefulness of electron spin resonance in detecting radical species.
- Give examples of photodissociation, ligand exchange and redox processes of d-block complexes.

Principles and Applications of Photochemistry   Brian Wardle
© 2009 John Wiley & Sons, Ltd

- Describe the principles of the use of $Ru(bpy)_3^{2+}$ as a sensitiser in artificial photosynthesis systems.
- Explain the photochemistry of organometallic compounds from the point of view of photodissociation of simple organometallics, metal-to-metal bond cleavage and the photosubstitution of cluster species.

7.1 THE PATHWAY OF PHOTOCHEMICAL REACTIONS

Since all photochemical reactions require the absorption of a photon, the result is that the reactant molecule is raised to a higher energy level. The outcome of this process depends on the nature of the upper and lower electronic states of the molecule. Four types of absorption behaviour are possible and we will first illustrate these by referring to Morse curves for the simple, diatomic, molecules. Although the potential energy of a complex molecule as a function of its molecular geometry is not a simple two-dimensional curve but a complex multidimensional surface, the conclusions arrived at by the use of Morse curves are instructive.

- In Figure 7.1, the electronic transition from lower to upper state results in promotion of the molecule to an upper vibrational level of the upper electronic state.

$$R(v=0) + h\nu \rightarrow R^*(v=n)$$

- In Figure 7.2 the transition indicated results in the excited molecule having potential energy greater than the maximum vibrational level, known as the bond dissociation energy (shown with a dotted line). This means that after absorption, the bond in the molecule will cleave at its first vibration.
- In Figure 7.3, the upper electronic state, known as a **dissociative state**, is unstable (it has no minimum) and has no vibrational levels. The electronic transition is always accompanied by bond cleavage.
- A further route to bond cleavage exists, as shown in Figure 7.4, utilising two upper electronic states that are close in energy, one of which is stable and one of which is unstable. Transition occurs from the lower state to a stable upper state. However, at

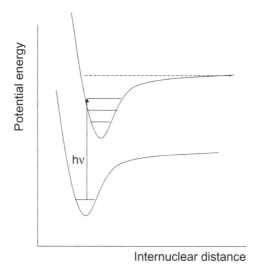

Figure 7.1 Photon absorption leading to a vibronic transition

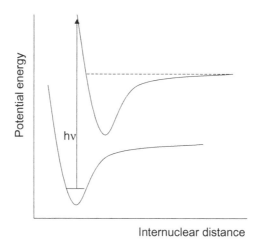

Figure 7.2 Photon absorption leading to a bound excited state with energy greater than the bond dissociation energy, resulting in bond cleavage

the point at which the two upper state curves cross, cleavage occurs as, during the course of a vibration, the excited molecule is changed to the unstable state. Such behaviour is known as **predissociation.**

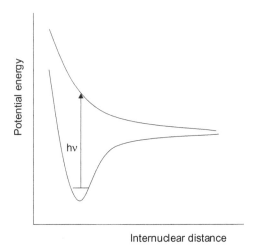

Figure 7.3   Absorption to a dissociative state results in bond cleavage

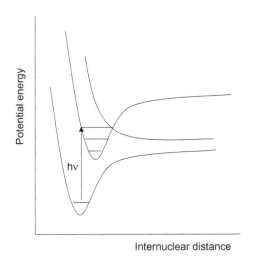

Figure 7.4   Promotion to a stable state followed by crossing to a dissociative state leads to cleavage by a process called predissociation

Another way of visualising the potential energy of a molecular system is in terms of the potential-energy changes that occur during a chemical reaction. This is represented by means of a reaction profile where the potential energy values of the reactants, products, transition states and intermediates are plotted against the reaction coordinate (the lowest

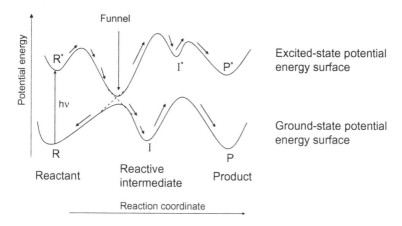

Figure 7.5 Pictorial representation of potential energy surfaces for excited-state and ground-state molecules. The arrows represent the course of a molecule moving along the reaction coordinate
Adapted from N. J. Turro, *Modern Molecular Photochemistry*, (1991). ©1991 University Science Books

energy pathway for the reaction). Rather than a single energy surface, which is adequate to model thermal, ground-state, reactions, modelling photochemical reactions requires at least two energy surfaces; that is, the ground-state surface and an excited-state surface from which the primary photochemical processes of the excited state, R*, occur.

We can approximate simple potential energy surfaces for the reacting ground-state and excited-state molecules in the manner shown in Figure 7.5.

All photochemical reactions that begin and end on the ground-state potential surface of a system involve crossing from an excited state to the ground state. This photochemical **funnel** occurs at the molecular structure, where the excited-state reactant or intermediate is delivered to the ground state to initiate the product formation. For many organic reactions, the structure of the funnel can take the form of a **conical intersection (CI)**, so called because in the region where the crossing occurs the equations for the two potential energy surfaces resemble the geometrical shape of two touching (or almost touching) cones.

Absorption of a photon by an organic molecule, R, leads to formation of an electronically-excited state, R*.

$$R + h\nu \rightarrow R^*$$

The primary photochemical process is concerned with the subsequent reactions of these electronically-excited states, which tend to react in one of two ways:

- A concerted (single-step) process, to give the product P:

$$R^* \rightarrow P$$

These include a series of pericyclic reactions initiated from $S_1(\pi,\pi^*)$. Pericyclic reactions are concerted reactions with a cyclic transition state. While in this transition state, a concerted rearrangement of electrons takes place, which causes σ- and π-bonds to simultaneously break and form.
- The formation of a reactive intermediate (I) or reactive intermediates

$$R^* \rightarrow I$$

These include the reactions of ketones initiated from $S_1(n,\pi^*)$ or $T_1(n,\pi^*)$ and proceeding via radical intermediates.

The secondary processes in photochemical reactions (the so-called dark processes) occur from the reactive intermediates, $I \rightarrow P$. These include the reactions of radical species (see Section 7.3).

7.2 DIFFERENCES BETWEEN PHOTOCHEMICAL AND THERMAL REACTIONS

When a molecule is irradiated with a frequency that matches the energy difference between the ground and excited state, a photochemical reaction may occur. Photochemical reactions differ significantly from thermal reactions.

- Photochemical reactions are the reactions of excited-state molecules initiated by photon absorption whereas thermal reactions are the reactions of ground-state molecules usually initiated by heat. The energy of photoexcitation of molecules can be provided by photon absorption even at very low temperatures and is of the same order as the activation energies for ground-state molecules. Provided the process of photoexcitation can be utilised in order to

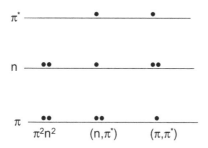

Figure 7.6 Electronic configurations of the ground-state and excited-state methanal molecules

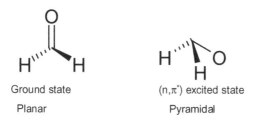

Figure 7.7 Molecular geometry of ground-state and (n,π*) excited-state methanal molecules

overcome the energy barrier of the ground-state molecule, the excited molecules will react more rapidly than the ground-state molecules.

- Photoexcited molecules have a different electronic distribution and a different molecular geometry than the corresponding ground-state molecules. If we consider methanal, $H_2C=O$, the lowest-energy electronic transition brought about by photon absorption involves an electron in the n molecular orbital (localised on oxygen) being promoted to the π^* molecular orbital (located between carbon and oxygen) (Figure 7.6).

Since the molecular geometry is related to electron distribution within the molecule, the ground-state and (n,π*) excited-state molecules have different geometries (Figure 7.7). The C-O bond length of the S_1 and T_1 excited states is longer than that of S_0, due to the excited states having less double-bonded character than S_0.

Excited-state molecules are more reactive than the corresponding ground-state molecules because:

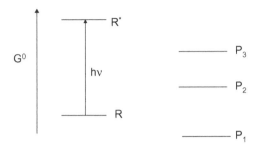

Figure 7.8 Gibbs-free-energy diagram showing a wider range of products accessible from the excited-state R* than from the ground state

- The bonds in R* include electrons in antibonding molecular orbitals.
- The bonds in R* are longer and weaker than those in R.
- There are a greater number of potential reactions accessible from the excited state than from the ground state (Figure 7.8).

    Thermodynamically-favourable reactions involve a decrease in Gibbs free energy, ΔG_r^0, for the reaction. Thus, potentially more products (P_1, P_2 and P_3) are available via the photochemical route than via the thermal route (P_1).

    Whereas a thermal reaction from R to P_2 or P_3 involves an increase in Gibbs free energy and so will not occur spontaneously, the photochemical reactions from R* involve a decrease in free energy and so are more likely to occur spontaneously. This allows the photochemical production of so-called 'energy-rich' or strained compounds such as P_2 and P_3 to be carried out at low temperatures, as these products may undergo decomposition at higher temperatures.

- In thermal reactions heat is applied to reactants, reaction media and products in an indiscriminate manner. In photochemical reactions a high concentration of excited species can be produced selectively by using monochromatic light of the correct energy at low temperature to produce monoenergetic products.
- Irradiation of a compound may produce different products than do thermal reactions. In a thermal reaction the chemical change favours the more thermodynamically-stable compound but the fate of a photochemical reaction is governed by using a wavelength that is absorbed by the reactant but not by the product. Thus the reaction pathways for excited-state species and ground-state species

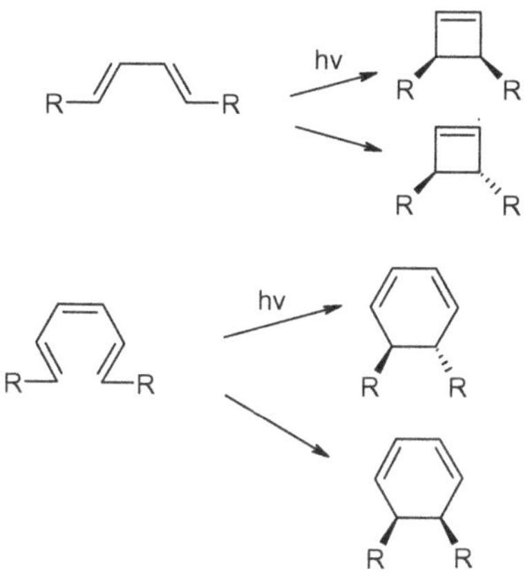

Figure 7.9 Electrocyclic processes of a diene and triene by photochemical and thermal pathways

can be very different, as can the stereochemical courses of the reactions.

For so-called electrocyclic processes, which are pericyclic reactions, the photochemical and thermal reactions give different stereoisomers, as shown for the diene and the triene in Figure 7.9.

The rules relating to which products are formed in these reactions are given in Section 8.3.

• Excited-state species are better electron donors and electron acceptors than the ground-state species. Thus the redox properties of the ground and excited states are different.

7.3 PHOTOLYSIS

When a molecule absorbs a quantum of light, it is promoted to an excited state. Because the energy of visible and ultraviolet light is of the same order of magnitude as that of covalent bonds, another possibility is that the molecule may cleave into two parts, a process known as **photolysis.**

One of the outcomes of this photolysis is the formation of **radicals;** that is, species containing unpaired electrons. Radicals are formed by homolytic fission of a covalent bond, where the electron pair constituting the bond is redistributed such that one electron is transferred to each of the two atoms originally joined by the bond:

$$\text{Br} \underset{}{\overset{\frown\frown}{\rule{3cm}{0.4pt}}} \text{Br} \xrightarrow{\;h\nu\;} \text{Br}\bullet + \bullet\text{Br}$$

Notice the use of 'fish-hook' arrows to show the movement of a single electron. Such arrows are used a great deal when writing mechanisms for reactions involving radical species.

Photochemically-generated radicals are encountered as reactive intermediates in many important systems, being a major driving force in the photochemistry of ozone in the upper atmosphere (stratosphere) and the polluted lower atmosphere (troposphere). The photochemistry of organic carbonyl compounds is dominated by radical chemistry (Chapter 9). Photoinitiators are used to form radicals used as intermediates in the chain growth and cross-linking of polymers involved in the production of electronic circuitry and in dental treatment.

7.3.1 Photohalogenation of Hydrocarbons

Irradiation of mixtures of halogen and hydrocarbons at suitable wavelengths can lead to halogenation of the hydrocarbon by a **free-radical chain reaction**.

1. Initiation:

$$\text{Br}_2 + h\nu \rightarrow 2\text{Br}\cdot$$

2. Propagation:

$$\text{Br}\cdot + \text{RH} \rightarrow \text{HBr} + \text{R}\cdot$$
$$\text{R}\cdot + \text{Br}_2 \rightarrow \text{RBr} + \text{Br}\cdot$$

3. Termination:

$$\text{Br}\cdot + \text{Br}\cdot \rightarrow \text{Br}_2$$
$$\cdot + \text{R}\cdot \rightarrow \text{R}_2$$
$$\text{R}\cdot + \text{Br}\cdot \rightarrow \text{RBr}$$

The initiation process produces the radical species Br•. Once formed, the Br• radical is consumed and regenerated in a repetitive cycle known as the propagation reactions (Figure 7.10).

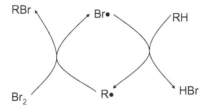

Figure 7.10 Repetitive nature of the propagation cycle in photobromination of a hydrocarbon (RH) by a free-radical chain reaction

The sum of the propagation reactions is equal to the overall stoichiometric equation of the reaction:

$$\begin{aligned}
Br\cdot + RH &\rightarrow R\cdot + HBr \\
R\cdot + Br_2 &\rightarrow RBr + Br\cdot \\
\hline
RH + Br_2 &\rightarrow RBr + HBr
\end{aligned}$$

Although the propagation reactions are only shown once, you should be aware that they occur in a sequence a very large number of times before the termination reactions remove the reactive radicals. Thus, free-radical chain reactions are characterised by the formation of a very large number of product molecules initiated by the absorption of a single photon in the initiation step; that is, chain reactions act as chemical amplifiers of the initial absorption step.

7.3.2 The Stratospheric Ozone Layer: Its Photochemical Formation and Degradation

Located several kilometres above the Earth's surface is the stratosphere. Here the ozone layer acts as a filter, protecting life on Earth from harmful low-wavelength ultraviolet radiation known as UV-C, which damages biological macromolecules such as proteins and DNA. In order to understand the effects of anthropogenic input into the stratosphere, the production and destruction of the ozone layer has been studied by a variety of photochemical models and experimental methods.

The simplest model for considering the stratospheric ozone layer is the **Chapman oxygen-only mechanism** (Figure 7.11), which describes the reactions' steady-state ozone concentration as resulting from a

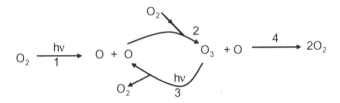

Figure 7.11 The Chapman cycle for the creation and destruction of odd oxygen in the stratosphere

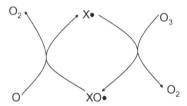

Figure 7.12 A catalytic cycle showing the loss of odd oxygen, where X• is a naturally-occurring radical (NO•, HO•, Cl• or Br•)

dynamic balance between the creation and destruction of so-called **odd oxygen** (O and O_3):

1. $O_2 + h\nu \rightarrow 2O$ (slow photolysis)
2. $O + O_2 \rightarrow O_3$ (fast exothermic)
3. $O_3 + h\nu \rightarrow O + O_2$ (fast photolysis)
4. $O + O_3 \rightarrow 2\,O_2$ (slow exothermic).

Although the Chapman mechanism models the general shape of the profile of stratospheric ozone correctly, it seriously overestimates its concentration. To account for this discrepancy, a catalytic process which increases the rate of step 4 occurs (Figure 7.12). Trace amounts of radicals present in the stratosphere reduce natural ozone levels below those predicted by the Chapman scheme:

$$X\cdot + O_3 \rightarrow XO\cdot + O_2$$
$$\underline{XO\cdot + O \rightarrow X\cdot + O_2}$$
$$O + O_3 \rightarrow 2O_2$$

Natural levels of catalytic radicals have been enhanced by the widespread use of chlorofluorocarbons (CFCs) and halons (e.g. CH_3Br). This has resulted in a massive loss of stratospheric ozone, particularly over the polar regions – the so-called ozone hole. This phenomenon is linked with the presence of chlorine and bromine atoms in the stratosphere, chiefly derived by the photochemical breakdown (photolysis) of the CFCs and halons.

$$CFCl_3 \xrightarrow{\ h\nu\ } \overset{.}{C}FCl_2 + \overset{.}{C}l$$
$$CH_3Br \xrightarrow{\ h\nu\ } \overset{.}{C}H_3 + \cdot Br$$

Reservoir molecules are stable molecules formed by reaction between catalytic radical species. They effectively 'lock up' the radicals and remove their ozone-destroying potential. However, in the polar spring, conditions are suitable for the formation of polar stratospheric clouds, on which fast heterogeneous chemistry occurs. These reactions release Cl_2 from reservoir molecules and this undergoes photolysis to ozone-destroying $Cl\bullet$ radicals, even in weak sunlight. Furthermore, nitric acid gets locked up in the clouds, thus reducing gas-phase concentrations of $\bullet NO_2$, which normally removes $ClO\bullet$ as $ClONO_2$.

Exposure of the air to sunlight brings about release of $Cl\bullet$ and $ClO\bullet$. Rapid ozone loss occurs via the cycle shown in Figure 7.13.

The international treaty known as the Montreal Protocol was signed in 1987, resulting in the phasing out of CFCs and other synthetic

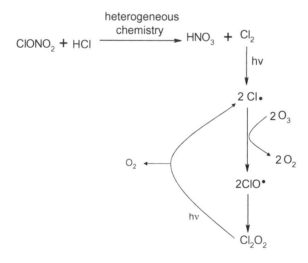

Figure 7.13 How heterogeneous reactions on polar stratospheric clouds result in rapid ozone loss

halocarbons. Total chlorine loading is now past its peak but recovery is occurring slowly due to the long lifetime of CFCs in the atmosphere.

7.3.3   Radicals in the Polluted Troposphere

Photochemical smog is the product of photochemical reactions caused by solar radiation and occurring in polluted air. The primary pollutants are nitrogen monoxide and hydrocarbons in the presence of bright sunlight. These pollutants generally arise from emissions from internal combustion engines and industrial plants. In extreme conditions, such as those found in Los Angeles, the primary pollutants are prevented from dispersing by a temperature inversion (lighter warm air lying on top of denser, cooler air).

Complex reactions involving radicals occur, giving rise to secondary pollutants such as ozone, aldehydes, peroxyacetyl nitrate (PAN) and particulate matter.

Catalytic hydroxyl radicals dominate the chemistry of the troposphere. These are formed by the following processes:

- Ultraviolet photolysis of ozone (there is a small background level of ozone in the troposphere as a result of downward transport from the stratosphere):

$$O_3 + h\nu \rightarrow O + O_2$$

- Reaction of O with water molecules:

$$O + H_2O \rightarrow 2 \cdot OH$$

- Reaction of hydroxyl radicals with hydrocarbons:

$$\cdot OH + RCH_3 \rightarrow RCH_2 \cdot + H_2O$$
$$RCH_2 \cdot + O_2 \rightarrow RCH_2O_2 \cdot$$
$$RCH_2O_2 \cdot + NO \cdot \rightarrow RCH_2O \cdot + NO_2 \cdot$$
$$RCH_2O \cdot + O_2 \rightarrow RCHO + HO_2 \cdot$$
$$HO_2 \cdot + NO \cdot \rightarrow \cdot OH + NO_2 \cdot$$
$$\overline{Net: RCH_3 + 2NO \cdot + 2O_2 \rightarrow RCHO + 2NO_2 \cdot + H_2O}$$

- Photolysis of the NO_2, producing O, which then forms ozone by reaction with O_2:

$$NO_2 \cdot + h\upsilon \rightarrow NO \cdot + O$$

$$O_2 + O \rightarrow O_3$$

Concentrations of O_3 and NO_2 may be so high that the ozone can be detected by its smell and a brown haze of NO_2 is clearly seen. The effect of the ozone is to damage vegetation, attack rubber and produce a marked increase in respiratory problems. In Los Angeles, ozone contributes to the formation of aerosols of particulate matter by its reaction with oils such as those found in pine trees and citrus fruits. This produces a hazy smokiness, reducing visibility. The particles also contain organic compounds known to be mutagens and carcinogens, which may be converted into much more potent mutagens by reaction with O_3 and NO_2

- Attack of OH on aldehydes, leading to formation of RCO radicals. These react with O_2 to yield peroxyacyl radicals which in turn react with NO_2 to form peroxyacyl nitrates:

$$\cdot OH + RCHO \rightarrow RCO\cdot + H_2O$$
$$RCO\cdot + O_2 \rightarrow RCOO_2\cdot$$
$$RCOO_2\cdot + NO_2\cdot \rightarrow RCOO_2NO_2$$

PAN, $CH_3COO_2NO_2$, is an important constituent of photochemical smog, acting as an irritant to the eyes and respiratory system and being highly toxic to plants.

7.4 AN INTRODUCTION TO THE CHEMISTRY OF CARBON-CENTRED RADICALS

Most stable ground-state molecules contain closed-shell electron configurations with a completely filled valence shell in which all molecular orbitals are doubly occupied or empty. Radicals, on the other hand, have an odd number of electrons and are therefore paramagnetic species. Electron paramagnetic resonance (EPR), sometimes called electron spin resonance (ESR), is a spectroscopic technique used to study species with one or more unpaired electrons, such as those found in free radicals, triplets (in the solid phase) and some inorganic complexes of transition-metal ions.

Free-radical chemistry is important in organic photochemistry as, generally speaking, all organic photochemical reactions of $^1(n,\pi^*)$ or $^3(n,\pi^*)$ excited states and all reactions of $^3(\pi,\pi^*)$ excited states result in the formation of either a pair of radicals or a biradical. We shall consider the basic principles relating to radical chemistry, allowing us to understand how radical and biradical intermediates react to form products.

The important reactions of radicals to be considered are:

- **Radical fragmentation** of radicals containing the carbonyl group ($R\dot{C}O$):

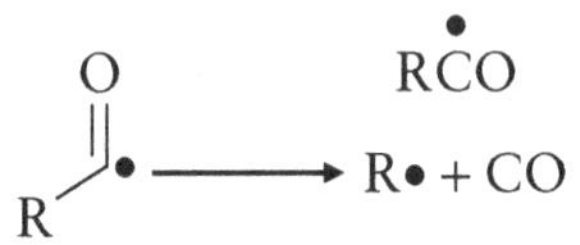

In the gas phase, this **decarbonylation** reaction is very efficient.

- **Radical rearrangements**, which are also possible if the change leads to a more stable product:

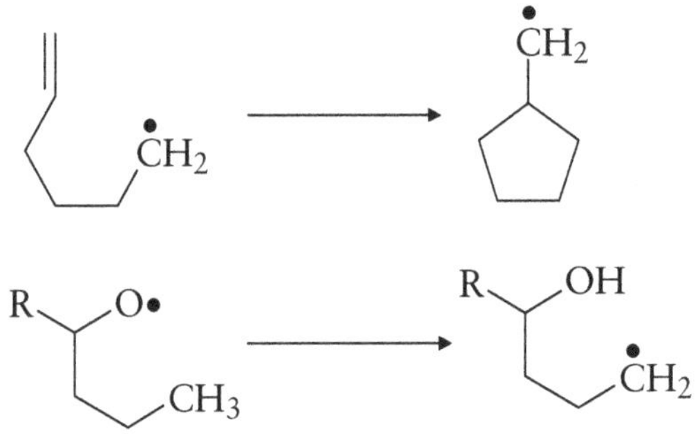

The overall process occurring in the above reactions is one of **intramolecular hydrogen abstraction.**

- Reactions between radicals resulting in **radical combination** or **disproportionation:**

$$\bullet C_2H_5 + \bullet C_2H_5 \rightarrow CH_3CH_2CH_2CH_3 \quad \text{Radical combination}$$

$$\bullet C_2H_5 + \bullet C_2H_5 \rightarrow C_2H_4 + C_2H_6 \quad \text{Disproportionation}$$

$$2\ \bullet C(CH_3)_3 \rightarrow (CH_3)_3C\text{-}C(CH_3)_3 \quad \text{Radical combination}$$

$$2\ \bullet C(CH_3)_3 \rightarrow (CH_3)_3CH + (CH_3)_2C{=}CH_2 \quad \text{Disproportionation}$$

Note that the disproportionation reaction involves **intermolecular hydrogen abstraction**. Due to steric factors, tertiary alkyl radicals undergo disproportionation rather than radical combination.

- Radicals forming addition products with compounds containing C=C bonds:

$$CH_3\dot{C}H_2 + PhCH=CH_2 \rightarrow CH_3CH_2CH_2\dot{C}HPh$$

Note that the product is a radical which can further react with $PhCH=CH_2$, leading to the process of vinyl polymerisation.

7.5 PHOTOCHEMISTRY OF THE COMPLEXES AND ORGANOMETALLIC COMPOUNDS OF d-BLOCK ELEMENTS

7.5.1 The Photochemistry of Metal Complexes

The principal photochemical reactions of metal complexes include dissociation, ligand exchange and redox processes. Unlike organic photoreactions (which take place almost exclusively from the S_1 or T_1 states), the excited state formed on irradiation depends on the wavelength employed. Hence the quantum yield often depends on the wavelength of the irradiating source. The excited-state processes give rise to a reactive intermediate which may find application in the synthesis of new compounds.

7.5.1.1 Dissociation

Consider irradiation of the complex:

$$[Cr(NH_3)_6]^3 + [Cr(NH_3)_6]^{3+} + h\nu \rightarrow [Cr(NH_3)_6]^{3+*} \rightarrow$$
$$[Cr(NH_3)_5]^{3+} + NH_3$$

Photoinduced isomerisation is sometimes possible where the metal-ligand bond undergoes homolysis followed by rapid combination of the metal and the rearranged ligand.

$$M\text{-}NO_2 + h\nu \rightarrow M\text{-}O\text{-}NO$$

For example:

$$[Co(NH_3)_5 NO_2] + h\nu \rightarrow [Co(NH_3)_5 NO_2]^* \rightarrow [Co(NH_3)_5 (ONO)]$$

7.5.1.2 Ligand exchange

A solvent molecule such as water will exchange for a ligand according to:

$$[Cr(NH_3)_6]^{3+} + h\nu \rightarrow [Cr(NH^3)_6]^{3+*}$$
$$[Cr(NH_3)_6]^{3+*} + H_2O \rightarrow [Cr(NH_3)_5 (H_2O)]^{3+} + NH_3$$

In mixed-ligand complexes, different products are often obtained in photochemical and thermal reactions.

Thermally, ammine complexes of chromium(III) containing a coordinated ligand X^- (where X^- is Cl^-, CNS^-, etc.) undergo substitution of X^- by H_2O in aqueous solution with retention of stereochemistry:

$$[Cr(NH_3)_5 (SCN)]^{2+} + H_2O \rightarrow [Cr(NH_3)_5 H_2O]^{3+} + SCN^-$$
$$\text{trans-}[Cr(NH_3)_4 Cl_2]^+ + H_2O \rightarrow \text{trans-}[Cr(NH_3)_4 (H_2O)Cl]^{2+} + Cl^-$$
$$\text{cis-}[Cr(NH_3)_4 Cl_2]^+ + H_2O \rightarrow \text{cis-}[Cr(NH_3)_4 (H_2O)Cl]^{2+} + Cl^-$$

In contrast, irradiation leads in many cases to different products and different stereochemistry:

$$[Cr(NH_3)_5 (SCN)]^{2+} + h\nu + H_2O \rightarrow [Cr(NH_3)_4 (SCN)(H_2O)]^{2+} + NH_3$$
$$\text{trans-}[Cr(NH_3)_4 Cl_2]^+ + h\nu + H_2O \rightarrow \text{cis-}[Cr(NH_3)_4 (H_2O)Cl]^{2+} + Cl^-$$

For ligand substitution and substitution-related (isomerisation and racemisation) reactions of complexes in solution, the difference between the thermal and photochemical reactions may be explained as:

- Thermally-induced substitutions occur via a pathway where the complex is in its electronic ground state.
- Irradiation of the complex results in population of an excited state whose energy and geometry may be different from those of the ground state.

7.5.1.3 Redox processes

Electron-transfer reactions have recently assumed great importance because of their potential use in artificial photosynthesis (see Section 12.4). The aims of artificial photosynthesis are to use excited-state electron transfer to produce an electrical potential and associated current, and to drive high-energy reactions such as the splitting of water into hydrogen and oxygen. The basis of the water-splitting reaction utilising photoredox chemistry is illustrated in Figure 7.14. The light-absorbing species acts as a photosensitiser and light absorption brings about an electron-transfer reaction between an electron donor, D, and acceptor, A, resulting in formation of energy-rich species D^+ and A^-. The oxidation of water to oxygen is brought about by D^+ mediated by a suitable catalyst such as RuO_2. Similarly, the catalytic reduction of water to hydrogen is brought about by A^- on a Pt catalyst.

The $Ru(bpy)_3^{2+}$ ion is a popular choice as a sensitiser for these photoredox reactions on account of it having the following properties:

- The complex has an intense MLCT absorption in the visible region.
- The lowest excited state, $Ru(bpy)_3^{2+}$, has a sufficiently long lifetime (~0.6 μs) to allow bimolecular reactions to compete with other deactivation processes.
- $Ru(bpy)_3^{2+}$ is both a stronger reductant and a stronger oxidant than $Ru(bpy)_3^{2+}$. Reductive quenching produces the powerful reductant

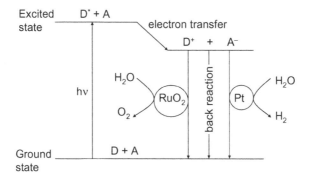

Figure 7.14 Photosensitised water-splitting reaction based on a photoredox reaction

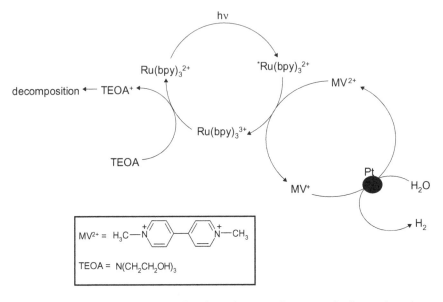

Figure 7.15   Photoredox system for the reduction of water to hydrogen based on reductive quenching of excited $^*Ru(bpy)_3^{2+}$ by methylviologen
Reprinted from C. Kutal, 'Photochemical Conversion and Storage of Solar Energy', *Journal of Chemical Education*, Volume 60 (10), 1983. © American Chemical Society

$Ru(bpy)_3^+$, while the powerful oxidant $Ru(bpy)_3^{3+}$ is produced by oxidative quenching.

- The coordinated 2,2'-bipyridyl ligands are inert to photosubstitution in aqueous solution.

Figure 7.15 shows a photoredox system capable of producing hydrogen from water. Photon absorption by $Ru(bpy)_3^{2+}$ results in formation of $^*Ru(bpy)_3^{2+}$, which undergoes oxidative quenching by methylviologen (MV^{2+}).

$$Ru(bpy)_3^{2+} + h\nu \rightarrow {}^*Ru(bpy)_3^{2+}$$
$$^*Ru(bpy)_3^{2+} + MV^{2+} \rightarrow Ru(bpy)_3^{3+} + MV^+$$

The rapid back-electron transfer between these products is deliberately impeded by reducing $Ru(bpy)_3^{3+}$ back to $Ru(bpy)_3^{2+}$ with the sacrificial triethanolamine (TEOA), and in the presence of colloidal platinum MV^+ reduces water to H_2.

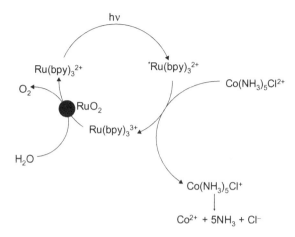

Figure 7.16 Photoredox system for the oxidation of water to oxygen based on oxidative quenching of excited $Ru(bpy)_3^{2+}$ by the sacrificial $Co(NH_3)_5Cl^{2+}$
Reprinted from C. Kutal, 'Photochemical Conversion and Storage of Solar Energy', *Journal of Chemical Education*, Volume 60 (10), 1983. © American Chemical Society

$$Ru(bpy)_3^{3+} + TEOA \rightarrow Ru(bpy)_3^{2+} + TEOA^+$$

$$MV^+ + H^+ \rightarrow MV^{2+} + \frac{1}{2}H_2$$

A photoredox system for the production of oxygen from water using $Ru(bpy)_3^{2+}$ as a sensitiser is shown in Figure 7.16. This system involves oxidative quenching of excited $Ru(bpy)_3^{2+}$ by the Co(III) complex $Co(NH_3)_5Cl^{2+}$:

$$*Ru(bpy)_3^{2+} + Co(NH_3)_5Cl^{2+} \rightarrow Ru(bpy)_3^{3+} + Co(NH_3)_5Cl^+$$

The Co(II) complex decomposes rapidly and irreversibly to Co^{2+}, which is so weak a reductant that it is unable to reduce the Ru(III) complex to Ru(II).

The $Ru(bpy)_3^{3+}$ then oxidises water to oxygen in a reaction mediated by the RuO_2 catalyst:

$$Ru(bpy)_3^{3+} + \frac{1}{2}H_2O \rightarrow Ru(bpy)_3^{3+} + \frac{1}{4}O_2 + H^+$$

7.5.2   An Aside: Redox Potentials Involved in Photoredox Reactions

The reduction potentials of some redox couples are given in Table 7.1. The potentials listed are written as if they are for reduction couples, with any reactions written as oxidation couples with potential values of the same magnitude but opposite sign.

The free-energy change, ΔG, per electron transferred related to the potential of the combined redox couples, E, is given by:

$$\boxed{\Delta G = -nFE}$$

where n is the number of electrons transferred and F is the Faraday.

Now, if a reaction is thermodynamically spontaneous in the direction it is written then ΔG for the reaction will have a negative sign.

That is:

Redox reactions that are thermodynamically spontaneous will have a positive value of E.

By referring to Table 7.1, it is possible to see why excited $^{*}Ru(bpy)_3^{2+}$ will reduce MV^{2+} but the ground-state species will not:

Table 7.1   Some reduction potentials for redox couples at pH 7

Couple	Potential/Volts
$Ru(bpy)_3^{2+}/Ru(bpy)_3^{+}$	−1.30
$Ru(bpy)_3^{3+}/^{*}Ru(bpy)_3^{2+}$	−0.83
MV^{2+}/MV^{+}	−0.45
H_2O/H_2	−0.42
$EDTA^{+}/EDTA$	0.20
$TEOA^{+}/TEOA$	0.80
O_2/H_2O	0.81
$Ru(bpy)_3^{3+}/Ru(bpy)_3^{2+}$	1.27
Co^{3+}/Co^{2+}	1.82

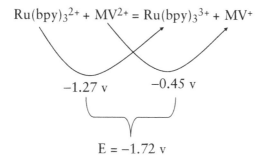

Thus the reaction of the ground-state Ru(II) species with MV^{2+} will not be thermodynamically spontaneous.

However, for the excited-state Ru(II) species:

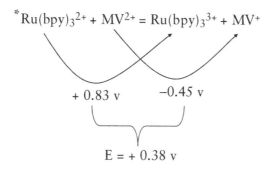

Since E has a positive value, ΔG has a positive sign, indicating that the reaction is thermodynamically spontaneous.

7.5.3   Organometallic Photochemistry

Organometallic compounds are those which contain at least one direct metal-to-carbon bond. The main classes of photochemical reaction of importance to organometallic compounds are given below.

7.5.3.1   Photosubstitution of metal carbonyls

The dominant photochemical reaction of metal carbonyl compounds is loss of carbon monoxide, which is usually followed by substitution of another ligand to replace the expelled carbon monoxide.

$$W(CO)_6 + PPh_3 \xrightarrow{hv} W(CO)_5(PPh_3) + CO$$
$$Fe(CO)_5 + 1,3\text{-butadiene} \xrightarrow{hv} Fe(CO)_3(1,3\text{-butadiene}) + 2CO$$

Preparative reactions involving photosubstitution of CO are carried out by irradiation of the compound in a weakly-coordinating solvent such as tetrahydrofuran (THF) and then displacement of the solvent ligand at room temperature with the chosen ligand, L.

$$Mn(\eta^5\text{-}C_5H_5)(CO)_3 + THF \xrightarrow[THF]{hv} Mn(\eta 5\text{-}C_5H_5)(CO)_2(THF) + CO$$
$$Mn(\eta^5\text{-}C_5H_5)(CO)_2(THF) + L \rightarrow Mn(\eta^5\text{-}C_5H_5)(CO)_2 L + THF$$

7.5.3.2 Photoinduced cleavage of metal-to-metal bonds

On photolysis, dinuclear single metal-to-metal-bonded carbonyls undergo homolysis of the metal-to-metal bond:

$$(CO)_5 Mn\text{-}Mn(CO)_5 \xrightarrow{hv} (CO)_5 Mn\cdot + \cdot Mn(CO)_5$$

The radical-like products readily abstract halide ligands from halo-carbon solvents:

$$Mn_2(CO)_{10} \xrightarrow[CCl_4]{hv} 2\, Mn(CO)_5 Cl$$

Photolysis of $W_2(\eta^5\text{-}C_5H_5)_2(CO)_6$ in the presence of Ph_3CCl results in W–W bond cleavage and the radical-like product abstracting Cl from Ph_3CCl, leaving $Ph_3C\bullet$ radicals, which can be positively identified by electron spin resonance (ESR).

$$(\eta^5\text{-}C_5H_5)(CO)_3 W\text{-}W(CO)_3(\eta^5\text{-}C_5H_5) \xrightarrow{hv} 2\cdot W(CO)_3(\eta^5\text{-}C_5H_5)$$
$$2\cdot W(CO)_3(\eta^5\text{-}C_5H_5) + 2Ph_3CCl \rightarrow 2\, ClW(CO)_3(\eta^5\text{-}C_5H_5) + 2Ph_3C\cdot$$

7.5.3.3 Photosubstitution of clusters

On irradiation, the cluster molecule $Os_3(CO)_{12}$ is resistant to Os–Os bond cleavage and instead undergoes photosubstitution with a low quantum yield:

$$Os_3(CO)_{12} + PPh_3 \xrightarrow{hv} Os_3(CO)_{11}(PPh_3) + CO$$

Generally, clusters of nuclearity >3 undergo photosubstitution with low quantum yield:

$$H_2FeRu_3(CO)_{13} + PPh_3 \xrightarrow{\ h\nu\ } H_2FeRu_3(CO)_{12}(PPh_3) + CO$$

8

The Photochemistry of Alkenes

AIMS AND OBJECTIVES

After you have completed your study of all the components of Chapter 8, you should be able to:

- Understand the importance and role of the lowest (π,π^*) excited state of alkenes in the photochemistry of alkenes.
- Describe the structure of the vertical and nonvertical excited states of alkenes and show how the interconversion and deactivation of these states leads to stereochemical isomerisation.
- Explain the dependence of the composition of the photostationary state on the wavelength of radiation used for irradiation.
- Compare and contrast the photosensitised isomerisation of alkenes with the method of direct irradiation.
- Outline the important features of the concerted photoreactions of C=C compounds, including the excited state involved, the stereochemistry and the factors which determine the direction of photochemical change.
- Explain the ideas relating to frontier orbitals when considering the stereochemistry of the electrocyclic reactions of dienes and trienes.
- When given a reaction sequence for a synthetic process, identify which photochemical processes are taking place and explain why the photochemical stages are useful in the synthetic sequence.

Principles and Applications of Photochemistry   Brian Wardle
© 2009 John Wiley & Sons, Ltd

- Describe the principal features of concerted and nonconcerted photocycloadditions of alkenes.
- Describe the photoaddition of water and alcohols with alkenes.
- Outline the importance the photochemistry of C=C bonds in the phototherapy of infants with neonatal jaundice, the process of vision and the effects of ultraviolet radiation on DNA.

8.1   EXCITED STATES OF ALKENES

Absorption of a photon by an alkene produces a (π,π^*) vertical (Franck–Condon) excited state in which the geometry of the ground state from which it was formed is retained. Since the (π,π^*) state has no net π bonding, there is little barrier to free rotation about the former double bond. Thus, relaxation takes place rapidly, giving a nonvertical (π,π^*) state with a lower energy and different geometry to the vertical excited state.

The relaxed nonvertical excited state is referred to as the **p-state** as it has adjacent p-orbitals which are orthogonal due to a 90° angle of twist from the geometry of the vertical excited state. Figure 8.1 shows that the same p-state is produced from both geometrical isomers, and similarly, rapid radiationless decay of this p-state can produce either the cis or the trans isomer.

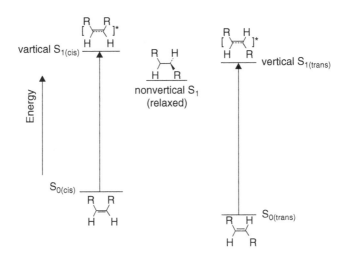

Figure 8.1   Energy diagram showing the vertical and nonvertical excited singlet states of an alkene

There is a very large energy difference between the $^1(\pi,\pi^*)$ and $^3(\pi,\pi^*)$ states for alkenes and so intersystem crossing is very inefficient and a triplet sensitiser is needed to populate the triplet state. Thus, different reaction conditions are required to form the excited singlet state (by direct irradiation) and the excited triplet state (by sensitised irradiation).

8.2 GEOMETRICAL ISOMERISATION BY DIRECT IRRADIATION OF C=C COMPOUNDS

When either the cis or the trans isomer of an alkene is irradiated, a mixture of both isomers will be formed in a particular ratio, which is dependent on the wavelength of light used. For example if either trans-stilbene or cis-stilbene is irradiated at 313 nm, the final composition of the reaction mixture will consist of a mixture of 93% cis and 7% trans (Figure 8.2).

This ratio is called the **photostationary state** composition. In the photostationary state, the rate of formation of each isomer from the nonvertical excited state is equal to its rate of removal by absorption of light. There is a roughly equal probability of the relaxation of the non-vertical excited state forming either the cis or the trans isomer and so the main factor influencing the photostationary state composition is the competition for absorbing light. This will, of course, depend on the relative values of the molar absorption coefficients of the two isomers at the particular wavelength chosen.

Consider the absorption spectra of the two geometrical isomers, shown in Figure 8.3. The photostationary state obtained on direct irradiation depends principally on the wavelength of light used, because of competitive absorption by the two ground-state isomers. If a wavelength is chosen that is absorbed more strongly by the trans isomer than by

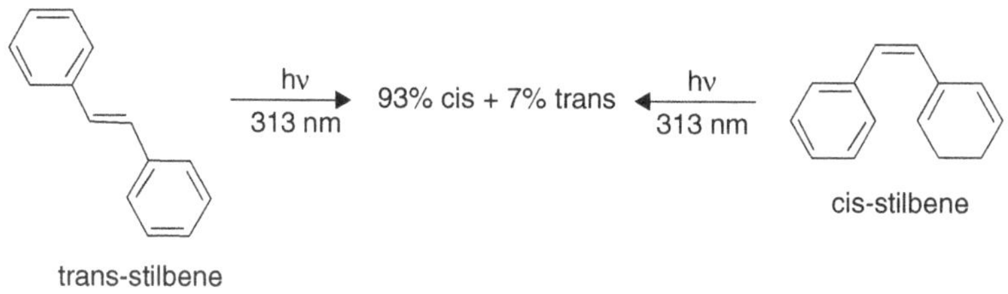

Figure 8.2   Product of irradiation of either trans-stilbene or cis-stilbene at 313 nm

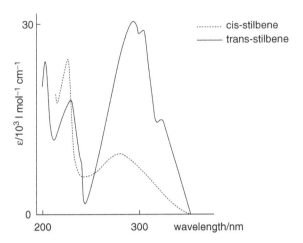

Figure 8.3   Absorption spectra of the geometrical isomers of stilbene

the cis isomer then the trans isomer will be excited preferentially and there will be less of the trans isomer in the photostationary state.

8.2.1   Phototherapy

Bilirubin is a product of the breakdown of haemoglobin in red blood cells. Neonatal jaundice occurs when bilirubin builds up faster than a newborn baby's liver is able to break it down. This results in the deposition of water-insoluble bilirubin in the skin (giving the skin a yellow colour) and untreated it can lead to damage of the central nervous system by deposition in brain cells.

In the bilirubin molecule there are two isomerisable double bonds, designated as $C_4=C_5$ and $C_{15}=C_{16}$, which normally exist as the cis,cis isomer (Figure 8.4).

On exposure to blue-green light, isomerisation of one or both double bonds takes place, forming trans,cis-bilirubin and trans,trans-bilirubin. In these compounds, hydrogen bonding to water molecules takes place, so that the molecule becomes increasingly water-soluble and can be excreted, relieving the baby of its toxic effect.

8.2.2   Vision

The initial act in the process of vision involves the photochemical cis-trans isomerisation of the 11-cis C=C bond of the retinal chromophore

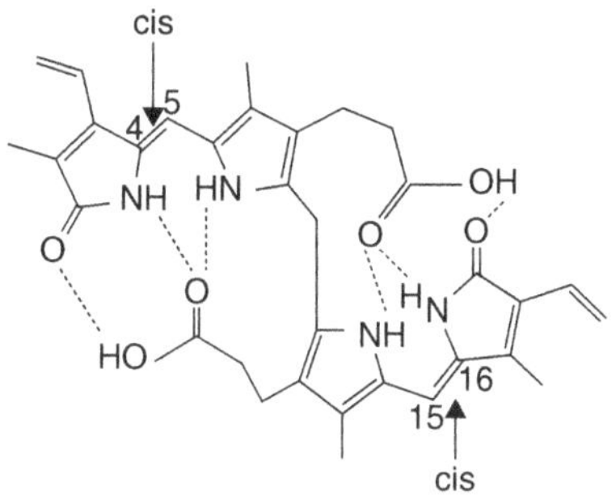

Figure 8.4 Cis,cis-bilirubin, showing intramolecular hydrogen bonds which produce a helical structure that is insoluble in water

in rhodopsin to form the all trans isomer. The retina of the eye is lined with many millions of photoreceptor cells, called rods and cones. The tops of the rods and cones contain a region filled with membrane-bound discs, which contain 11-cis-retinal bound to a protein called opsin. The resulting complex is called rhodopsin or 'visual purple'.

When visible light hits the cis-retinal, the 11-cis-retinal undergoes the process of cis–trans isomerisation to all-trans-retinal. The trans-retinal does not fit as well into the protein, and so a series of geometrical changes in the protein occur, initiating a cascade of biochemical reactions which results in a potential difference building up across the plasma membrane. This potential difference is passed along to an adjoining nerve cell as an electrical impulse. The nerve cell then carries the impulse to the brain, where the visual information is interpreted. The process of vision is dealt with in more detail in Section 12.3.

8.3 PHOTOSENSITISED GEOMETRICAL ISOMERISATION OF C=C COMPOUNDS

The geometrical isomerisation of alkenes can also occur through triplet excited states, $^3(\pi,\pi)$. A triplet sensitiser is used to absorb radiation, undergoing vibrational relaxation and intersystem crossing to the excited triplet state. The excited triplet state of the sensitiser then takes part in an energy-transfer reaction with the ground-state alkene. Transfer of energy in this manner excites the alkene to its triplet state, which can then decay to either geometrical isomer (Scheme 8.1).

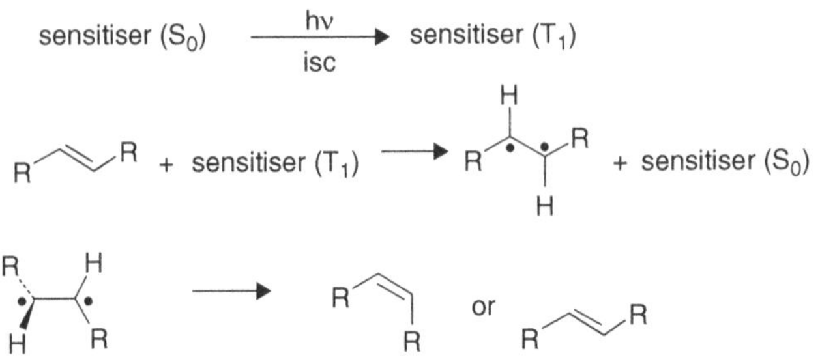

$$\text{sensitiser (S}_0\text{)} \quad \xrightarrow[\text{isc}]{h\nu} \quad \text{sensitiser (T}_1\text{)}$$

Scheme 8.1 Photosensitised isomerisation of alkenes

Unlike the singlet-state reactions, sensitised cis–trans interconversion is not dependent on the absorption coefficients of the two isomers at the wavelength of irradiation. The principal factor affecting the photostationary state is the triplet energy of the sensitiser compared to that of the isomeric alkenes. In order for the energy-transfer process to occur, the triplet energy of the sensitiser must be greater than that of the alkene. If the sensitiser has a triplet energy between those of the cis and trans isomers then the isomer with the lower triplet energy becomes preferentially sensitised and the isomer of higher triplet energy predominates in the photostationary state.

Consider the photosensitised cis–trans isomerisation of stilbene: The two isomers have different triplet energy levels ($247\,\mathrm{kJ\,mol^{-1}}$ for the cis and $205\,\mathrm{kJ\,mol^{-1}}$ for the trans). The proportion of the cis isomer in the photostationary state varies with the energy of the photosensitiser (E_T), as shown in Figure 8.5.

- When E_T(sensitiser) > E_T(cis), the rates for cis → trans and trans → cis conversion are similar and constant. Thus the proportion of the cis isomer is almost constant.
- As E_T(cis) > E_T(sensitiser) > E_T(trans), the rate of cis → trans becomes less than that of trans → cis. So more trans molecules are converted into cis and the percentage of the cis isomer increases.

8.3.1   Synthesis

The Wittig process for the synthesis of vitamin A acetate is carried out on the industrial scale and produces a mixture of the all-trans and 11-cis isomers. Only the all-trans form is suitable for pharmaceutical or

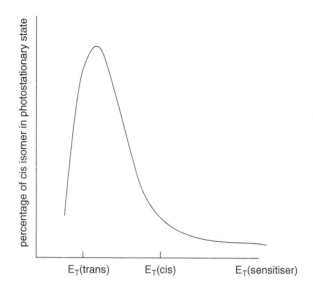

Figure 8.5 Effect of the sensitiser triplet energy on the percentage of cis-stilbene in the photostationary state
Adapted from W.G. Herkstroeter and G.S. Hammond, 'Energy transfer study by kinetic spectrophotometry', *Journal of the American Chemical Society*, Volume 88. © American Chemical Society

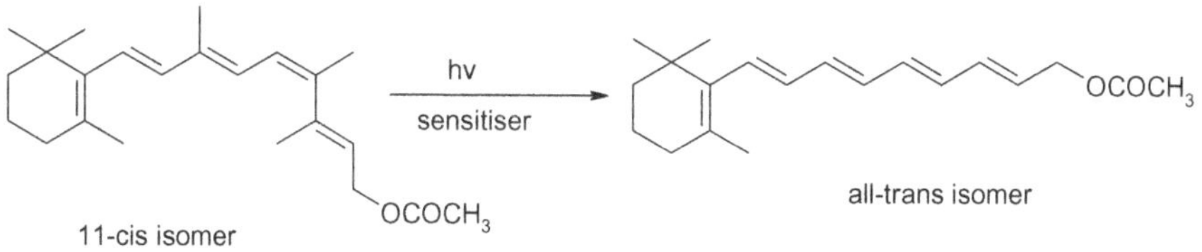

Scheme 8.2   Sensitised cis–trans isomerisation of vitamin A acetate

nutritional use and so a photochemical method has been developed in which the stereoisomeric mixture is irradiated with visible light in the presence of chlorophyll, which acts as the sensitiser (Scheme 8.2).

8.4   CONCERTED PHOTOREACTIONS

A concerted reaction is a chemical reaction in which all bond breaking and bond formation occurs in a single step in which reactive intermediates are not involved. Concerted photoreactions tend to be stereospecific, occurring from the vertical excited state.

8.4.1 Electrocyclic Reactions

In an electrocyclic ring-closure reaction a new σ-bond is formed between
the end atoms of a conjugated system of double bonds. Consider the
cyclisation of buta-1,3-diene:

This reaction can occur either thermally or photochemically, with the
equilibrium position (thermal reaction) depending on the thermodynamic stability of the reactant and product, and the photostationary-
state composition depending on the relative values of the absorption
coefficients at the wavelengths used.

Conjugated dienes can undergo photochemical conversion into
cyclobutenes:

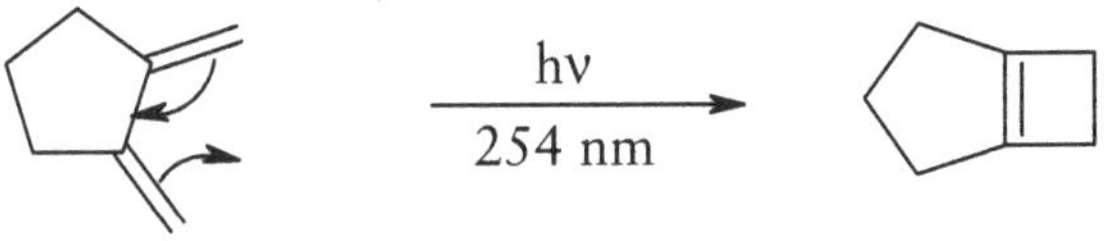

The reverse process may be possible, using short-wavelength radiation
in a region where the cyclobutene absorbs strongly:

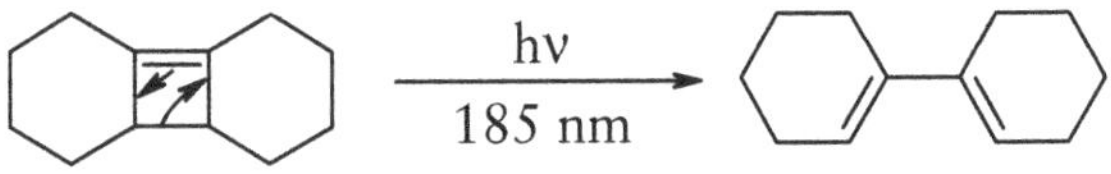

Thermal and photochemical electrocyclic reactions are both stereospecific, with the two processes giving rise to stereospecific reactions
in the opposite sense.

If we consider the two modes of ring closure in a 1,4-disubstituted
buta-1,3-diene, the two substituent groups in the cyclobutene have a
stereo-spatial arrangement that is brought about by either:

- a **conrotatory** mode of reaction in which rotations occur in the
  same sense about the axes of the C=C bonds:

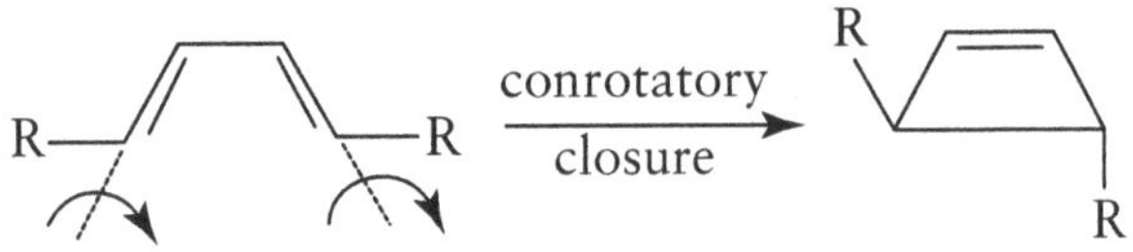

Table 8.1 Mode of reaction of electrocyclic reactions

N	Thermal Reaction	Photochemical Reaction
4n	conrotatory	disrotatory
4n + 2	disrotatory	conrotatory

or

- a **disrotatory** mode of reaction in which rotations occur in opposite senses about the axes of the C=C bonds:

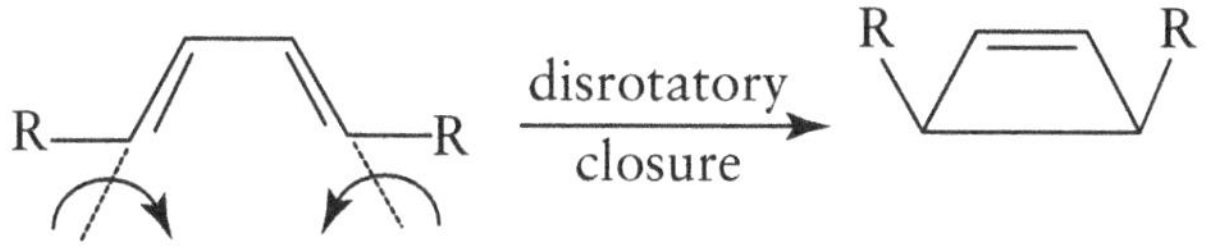

The mode of reaction of electrocyclic reactions is shown in Table 8.1, where the total number of electrons (N) is given as a multiple (4n) or not a multiple (4n + 2) of four.

In the simple four-electron systems, a route for cis–trans isomerisation of a diene is made available by the photochemical reaction usually being a disrotatory ring closure and the thermal reaction being a conrotatory ring opening:

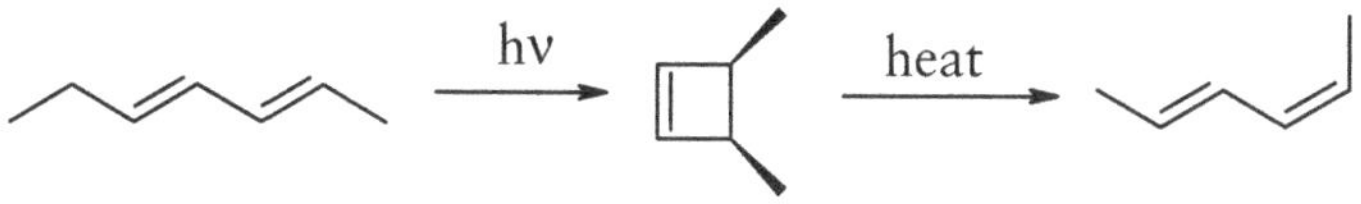

As an example of a six-electron system, trans-trans-octa-2,4,6- triene illustrates the opposite sense of the stereochemical nature of the thermal and photochemical reactions:

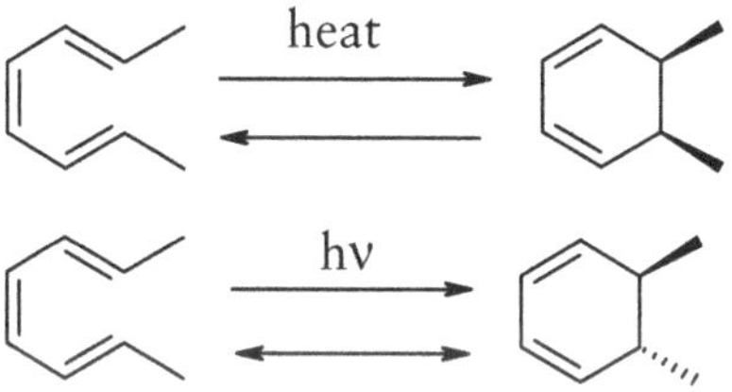

Cis-stilbene undergoes electrocyclic ring closure on irradiation:

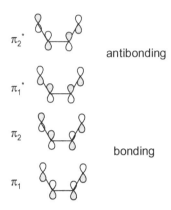

Figure 8.6 The four molecular orbitals derived from overlap of p atomic orbitals

The basis of electrocyclic reactions can be considered in terms of the molecular orbitals involved. By considering the phasing of the molecular orbitals it is possible to say whether a reaction can proceed (only orbitals of the same phase can overlap and bond) and to predict the stereochemistry of the reaction. This approach is called the **frontier orbital model**.

The molecular orbitals derived from overlap of four p atomic orbitals as found in dienes are shown in Figure 8.6.

For the thermal electrocyclic reaction of dienes, the HOMO for the diene is π_2, since there are four electrons to accommodate in the π-orbitals (two paired electrons per orbital). Thus, for hexa-2,4-diene the conrotatory mode of reaction gives the trans isomer (Scheme 8.3).

For the photochemical electrocyclic reaction of the diene, irradiation promotes one electron from π_2 to π_1^* and the disrotatory mode of reaction gives the cis isomer (Scheme 8.4).

The molecular orbitals derived from overlap of six p atomic orbitals as found in trienes are shown in Figure 8.7. Since there are six electrons to accommodate, as two paired electrons per orbital, the HOMO is π_3. Thus, for 2,4,6-octatriene the disrotatory mode of reaction gives the trans isomer (Scheme 8.5).

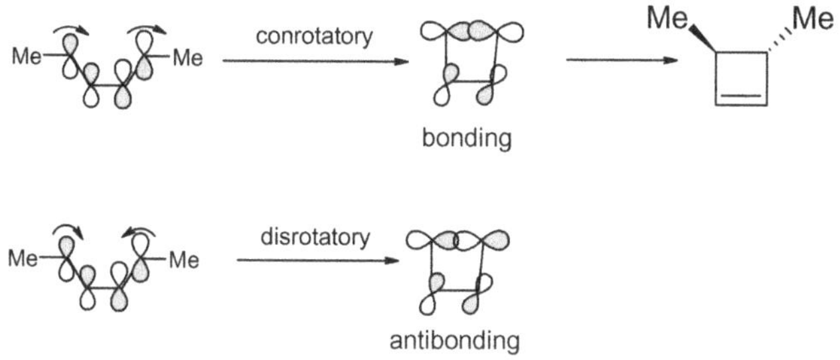

Scheme 8.3   Thermal electrocyclic reaction of a diene

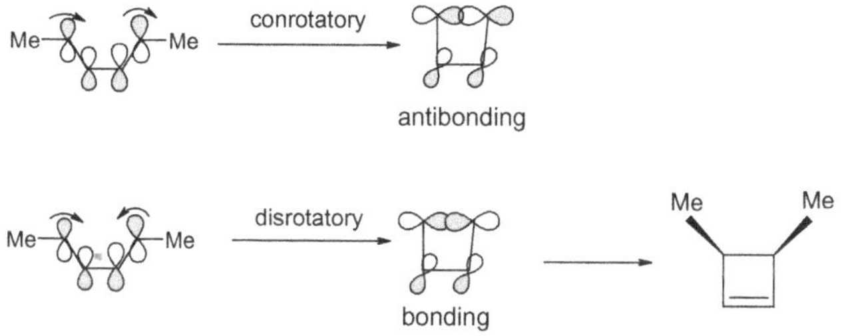

Scheme 8.4   Photochemical electrocyclic reaction of a diene

For the photochemical electrocyclic reaction of dienes, irradiation promotes one electron from π_3 to π_1^*. For 2,4,6-octatriene, the conrotatory mode of reaction gives the trans isomer (Scheme 8.6).

8.4.2   Sigmatropic Shifts

Sigmatropic shifts are reversible reactions in which a σ-bonded group (usually H) migrates from one point to another within a molecular system. Sigmatropic shifts may occur either thermally or photochemically, but 1,3- and 1,7-shifts are the most commonly encountered for the photochemical reactions.

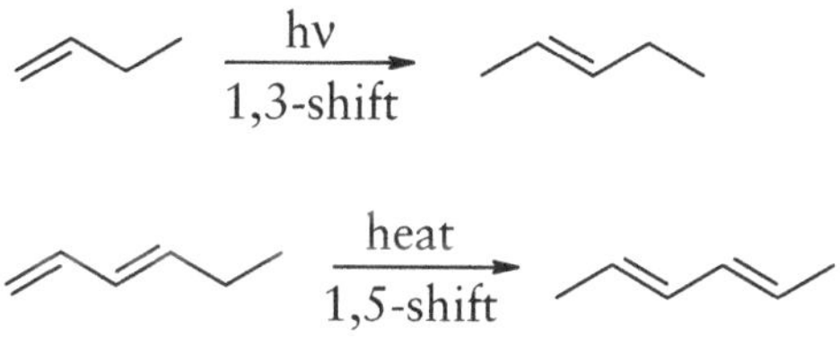

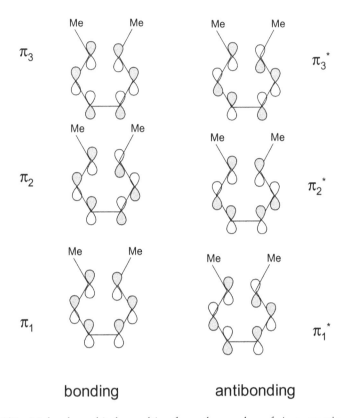

π_3 π_3^*

π_2 π_2^*

π_1 π_1^*

bonding   antibonding

Figure 8.7 Molecular orbitals resulting from the overlap of six p atomic orbitals. Note that the energy of $\pi_1 < \pi_2 < \pi_3 < \pi_1^* < \pi_2^* < \pi_3^*$

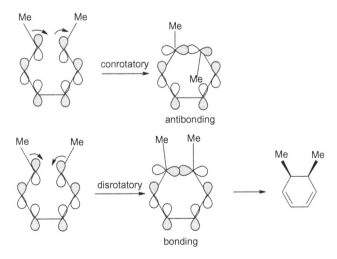

Scheme 8.5 Thermal electrocyclic reaction of a triene

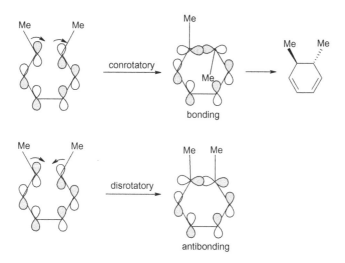

Scheme 8.6   Photochemical electrocyclic reaction of a triene

8.5   PHOTOCYCLOADDITION REACTIONS

A cycloaddition reaction produces a ring of atoms by forming two new σ-bonds, for example the formation of a cyclobutane dimer from two alkene molecules. The direct photoreaction involves the concerted reaction of the singlet $^1(\pi,\pi^*)$ excited state of one alkene with the ground state of the other. Stereospecific reactions in which the dimers preserve the ground-state geometry occur when liquid cis- or trans-but-2-ene are irradiated at low temperature:

For rigid alkenes, triplet sensitisation brings about photocycloaddition via the $^3(\pi,\pi^*)$ state. These reactions are neither concerted nor stereospecific. Cyclopentene produces a tricyclic dimer:

8.5.1   Solar Energy Storage

Triplet-sensitised photocycloaddition of norbornadiene (N) leads to formation of the energy-rich quadricyclane (Q) by an intramolecular reaction:

This reaction has the potential to be used in systems for the collection and storage of solar energy.

$$(\text{sensitiser})S_0 \xrightarrow{\text{isc}} (\text{sensitiser})T_1$$
$$(\text{sensitiser})T_1 + N \rightarrow (\text{sensitiser})S_0 + N(T_1)$$
$$N(T_1) \rightarrow Q$$

Although quadricyclane is highly strained, it is kinetically very stable on account of the large activation energy for the $Q \rightarrow N$ reaction. Use of suitable catalysts allows the stored solar energy to be released as heat.

When a C=C bond is conjugated with a C=O group, the carbonyl group lowers the energy absorption band of the compound so that it absorbs at longer (more accessible) wavelengths. Also, the process of intersystem crossing is enhanced, leading to formation of the excited triplet state on irradiation. For such triplet reactions, a mixture of stereoisomers is produced.

8.6 PHOTOADDITION REACTIONS

Ground-state alkenes generally undergo electrophilic addition with alcohols in the presence of a Brønsted acid catalyst, yielding the Markovnikov product:

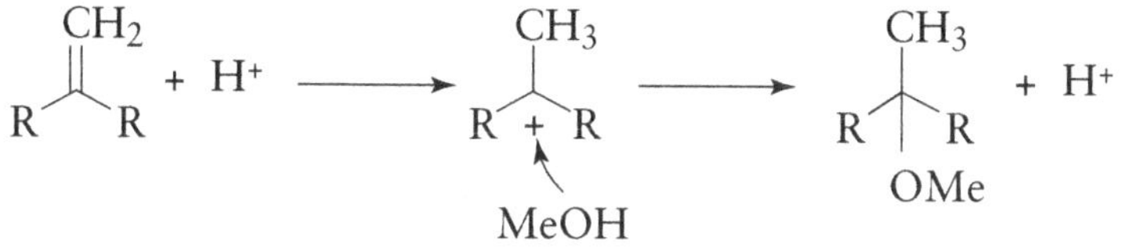

The excited states of alkenes can undergo sensitised addition reactions with simple molecules such as alcohols and water. The reactions are sensitised under mild conditions where acidic conditions are not required. Suitable sensitisers are:

- Electron-donor sensitisers such as 1-methoxynaphthalene (forming the Markovnikov addition product):

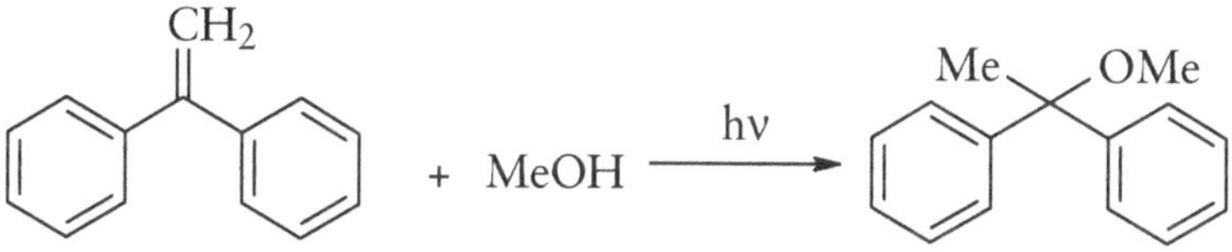

- Electron-acceptor sensitisers such as methyl p-cyanobenzoate (forming the anti-Markovnikov addition product):

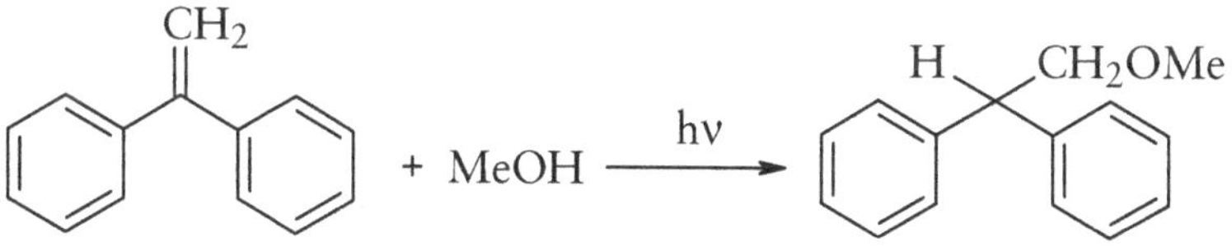

8.6.1 DNA Damage by UV

Photocycloaddition and photoaddition provide good models for the mechanism by which ultraviolet light can cause damage to nucleic acids. When ultraviolet light damages DNA, the principal reactions occurring involve the nucleic acid bases. Photocycloaddition reactions to form cyclobutane derivatives and photochemical hydration of C=C bonds to form addition products occur.

Thymine and other bases in nucleic acids undergo dimerisation when irradiated with ultraviolet light, and this can result in DNA mutation or even cell death.

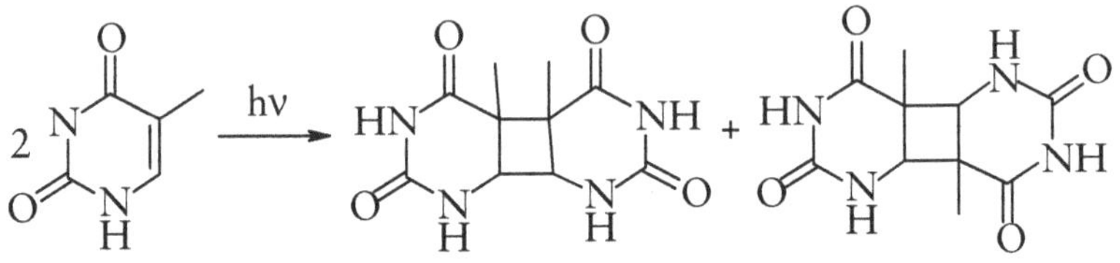

The cyclobutane derivatives can revert to the original bases when irradiated with shorter-wavelength light. This reaction is involved in the repair process, which helps keep damage caused by ultraviolet to a minimum. Some repair mechanisms involve enzymes that are important in the breakdown of cyclobutane dimers by longer-wavelength light.

Thymine also undergoes a photoaddition reaction with water to form the anti-Markovnikov product:

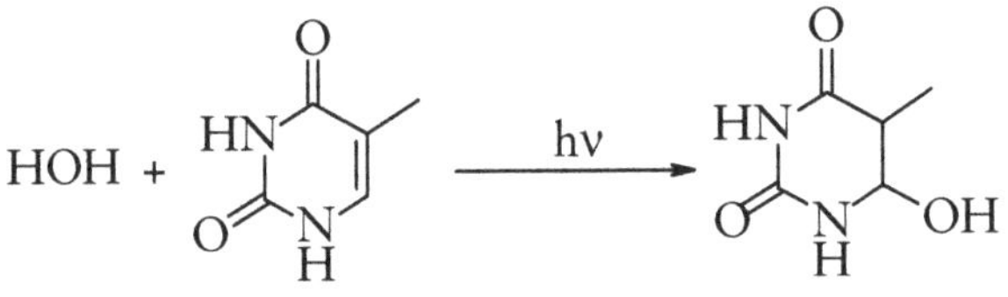

The reverse process, important in the repair process, occurs when the hydrate is irradiated with shorter-wavelength light.

9

The Photochemistry of Carbonyl Compounds

AIMS AND OBJECTIVES

After you have completed your study of all the components of Chapter 9, you should be able to:

- Predict the products of α-cleavage (Norrish type 1) reactions of given carbonyl compounds.
- Give examples of simple radical and biradical reactions – combination, disproportionation, hydrogen abstraction and fragmentation.
- Predict the products of the intermolecular hydrogen abstraction reactions of ketones and discuss the mechanisms of these reactions.
- Explain how intramolecular hydrogen abstraction in carbonyl compounds can lead either to cleavage (Norrish type 2 reaction) or to the formation of cyclic compounds (Yang cyclisation).
- Predict the products of the reactions of excited-state carbonyl compounds with alkenes (Paterno–Büchi reaction).
- Explain the role of the photochemical reactions of carbonyl compounds in the photoinitiated polymerisation of vinyl monomers and cross-linking in polymers.
- Discuss the role played by carbonyl compounds in the photodegradation (both unwanted and designed) of polymers.

Principles and Applications of Photochemistry Brian Wardle
© 2009 John Wiley & Sons, Ltd

9.1 EXCITED STATES OF CARBONYL COMPOUNDS

The aliphatic ketones have a weak absorption due to n → π* transition. Intersystem crossing from the (n,π*) singlet to triplet state is generally efficient because of their closeness in energy. The photochemistry of aliphatic ketones may involve singlet or triplet (or both) excited states. The unpaired electron in a nonbonding orbital on oxygen gives rise to the characteristic radical-like reactions of (n,π*) states.

Aromatic ketones have lowest $S_1(n,\pi^*)$ states and the (n,π*) and (π,π*) triplet states are close together in energy. Which of these is the lower in energy depends on the nature of the substituents on the aromatic rings and on external factors such as the nature of the solvent.

Electron-donating substituents in conjugation with the carbonyl group have the effect of stabilising the (π,π*) triplet state but destabilise the (n,π*) triplet state. On the other hand, the (n,π*) triplet state is stabilised relative to the (π,π*) triplet state by the presence of electron-accepting substituents.

Because an increase in solvent polarity results in a shift to longer wavelength of (π,π*) absorptions and longer wavelength of (n,π*) absorptions, polar solvents stabilise the (π,π*) triplet state relative to the (n,π*) triplet state. This effect may result in the (π,π*) triplet being the lowest triplet state.

Intersystem crossing in aromatic ketones is very efficient, so that their photochemistry is dominated by the triplet-state processes. The efficient formation of triplet states, the small singlet–triplet energy and the accessible long-wavelength absorption make the aryl ketones excellent triplet sensitisers.

The principal reaction types for ketone (n,π*) excited states are:

- α-cleavage; that is, cleavage of a C-R bond adjacent to the carbonyl group:

$$\underset{R}{\overset{R}{\diagdown}}C{=}O^* \longrightarrow R{-}\overset{\bullet}{C}{=}O + R\bullet$$

- Hydrogen abstraction from a suitable donor molecule R′H:

$$\underset{R}{\overset{R}{\diagdown}}C{=}O^*\,H{-}R' \longrightarrow \underset{R}{\overset{R}{\diagdown}}\overset{\bullet}{C}{-}OH + \bullet R'$$

- Addition to a C=C bond:

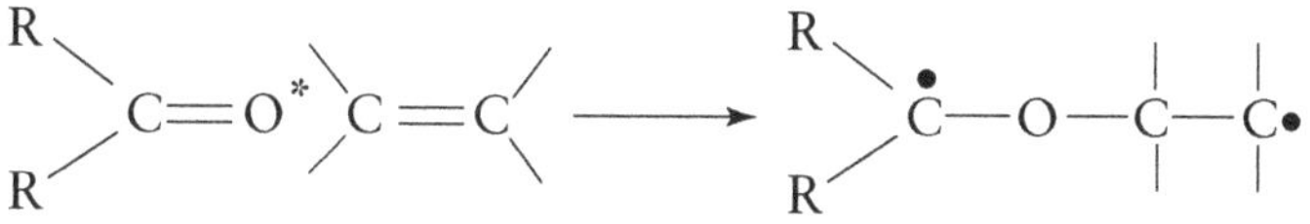

$$\underset{R}{\overset{R}{\diagdown}}C{=}O^* \quad \overset{}{\diagup}C{=}C\overset{}{\diagdown} \longrightarrow \underset{R}{\overset{R}{\diagdown}}\overset{\bullet}{C}{-}O{-}\overset{|}{\underset{|}{C}}{-}\overset{|}{\underset{|}{C}}\bullet$$

9.2 α-CLEAVAGE REACTIONS

α-cleavage (also known as Norrish Type 1 reaction) involves the cleavage of one of the bonds adjacent to the C=O group, leading initially to the production of radicals. With propanone (acetone), for example, methyl and acetyl radicals are formed:

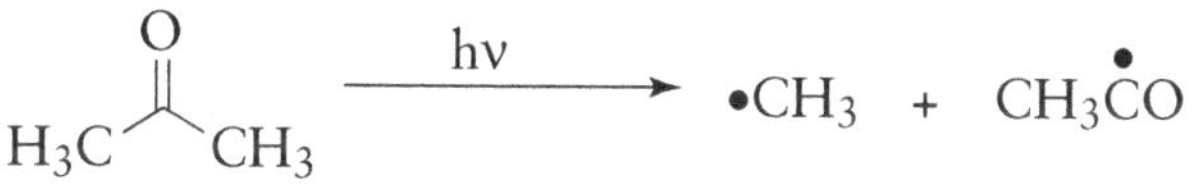

$$\underset{H_3C \quad CH_3}{\overset{O}{\overset{\|}{C}}} \xrightarrow{\quad h\nu \quad} \bullet CH_3 \; + \; CH_3\overset{\bullet}{C}O$$

In solution, diffusion apart of the radicals is inhibited by the solvent, resulting in the radicals recombining to form propanone. In the gas phase, however, the radicals do not combine and the acetyl radical breaks down to form carbon monoxide and another methyl radical. This elimination is known as **decarbonylation**. The methyl radicals then combine and the overall products are ethane and carbon monoxide (Scheme 9.1).

In the case of asymmetrical ketones, two different modes of α-cleavage can occur, with the major products being formed via the more stable pair of initially-formed radicals. For alkyl radicals, the stability of the radical increases as its complexity increases and radical stabilities are tertiary > secondary > primary.

$$CH_3\overset{\bullet}{C}O \longrightarrow \bullet CH_3 + CO$$

$$\bullet CH_3 + \bullet CH_3 \longrightarrow CH_3CH_3$$

$$\text{overall:} \quad CH_3COCH_3 \xrightarrow{\quad h\nu \quad} CH_3CH_3 + CO$$

Scheme 9.1

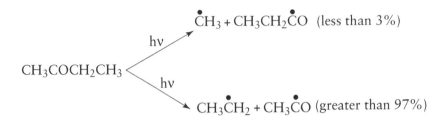

The efficiency of product formation in solution is also controlled by the stabilities of the radicals. Stable radicals such as tertiary alkyl radicals or benzyl radicals lead to efficient decarbonylation in solution. Because of steric factors involving bulky groups, tertiary radicals tend to preferentially undergo disproportionation rather than radical combination and so the quantum yield of the products formed by disproportionation exceeds that of the radical combination product.

$$(CH_3)_3C \cdot + (CH_3)_3C \cdot \rightarrow (CH_3)_3CH + (CH_3)_2C{=}CH_2$$

Thus, for the irradiation of $(CH_3)_3CO(CH_3)_3$ in solution at $313\,nm$:

$$(CH_3)_3CO(CH_3)_3 \xrightarrow{h\nu} CO + (CH_3)_3CH + (CH_3)_2C{=}CH_2 + (CH_3)_3C{-}C(CH_3)_3$$
$$\phi{=}0.6 \quad \phi{=}0.5 \quad \phi{=}0.5 \qquad \phi{=}0.1$$

disproportionation products

radical combination product

If radicals are unable to undergo disproportionation they must react by radical combination. For example, irradiation of $PhCH_2COCHPh_2$ gives three radical combination products in the statistically-expected ratio of $1:2:1$:

$$PhCH_2COCHPh_2 \xrightarrow{h\nu} PhCH_2CH_2Ph + PhCH_2CHPh_2 + Ph_2CHCHPh_2$$
$$25\% \qquad\qquad 50\% \qquad\qquad 25\%$$

If the free radicals are formed in equal amounts:

$$PhCH_2COCHPh_2 \xrightarrow{h\nu} PhCH_2\overset{\bullet}{C}O + \overset{\bullet}{C}HPh_2 \xrightarrow{-CO} Ph\overset{\bullet}{C}H_2 + Ph_2\overset{\bullet}{C}H$$

then the observed ratio is expected on the basis of equal encounter rates for any two radicals.

Cyclic ketones undergo decarbonylation on irradiation in the vapour phase and in solution, provided they form stabilised radicals:

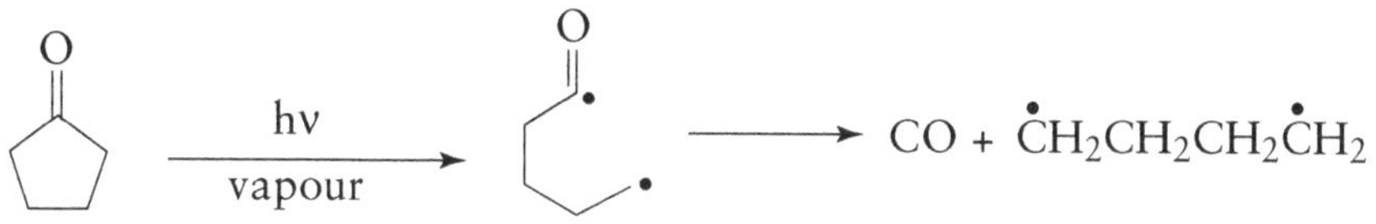

$$\text{CO} + \overset{\bullet}{\text{CH}_2}\text{CH}_2\text{CH}_2\overset{\bullet}{\text{CH}_2}$$

The biradical can undergo either cyclisation or bond cleavage:

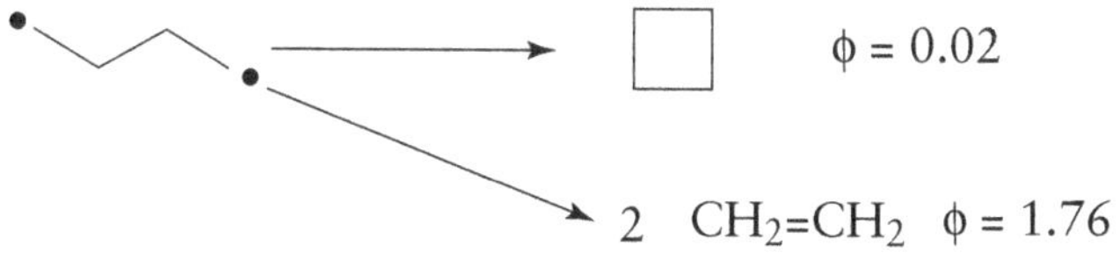

$\phi = 0.02$

$2 \ \text{CH}_2\text{=CH}_2 \ \ \phi = 1.76$

Five- or six-membered saturated cyclic ketones can also react by another pathway that does not involve decarbonylation. In these reactions, the biradical initially formed by α-cleavage undergoes internal disproportionation without loss of carbon monoxide, resulting in the formation of either an unsaturated aldehyde or a ketene. Methanol is usually added to convert the reactive ketene to a stable carboxylic-acid derivative (Scheme 9.2).

Photochemical α-cleavage of carboxylic acids results in loss of carbon dioxide (decarboxylation) rather than loss of carbon monoxide. The compound 2,4-dichlorophenoxyacetic acid, commonly known as (2,4-D), has been used extensively as a herbicide. This has posed a problem, because of the slow natural decomposition of 2,4-D in the environment. 2,4-D undergoes α-cleavage, undergoing decarboxylation:

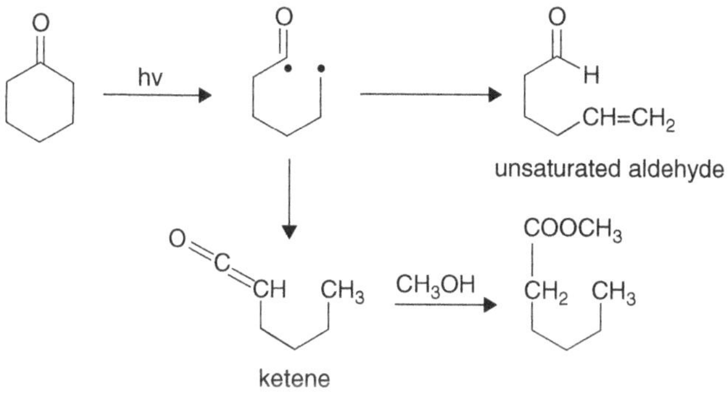

Scheme 9.2

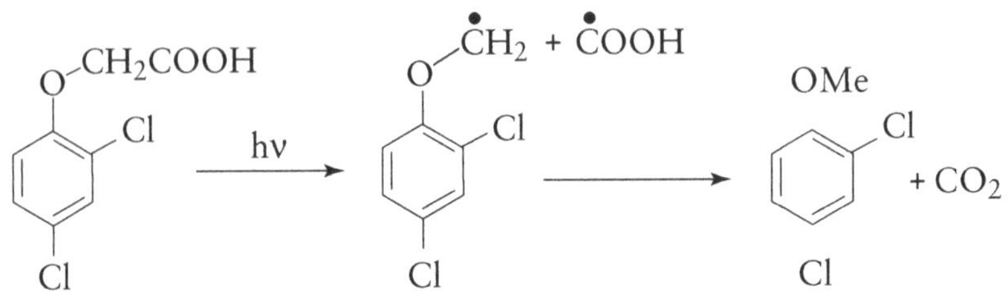

2,4-D

The photodecomposition is accelerated by using a sensitiser. If such a sensitiser is added to the herbicide it will result in a less environmentally-persistent herbicide.

9.3 INTERMOLECULAR HYDROGEN-ABSTRACTION REACTIONS

The second principal reaction of the (n,π*) excited state of carbonyl compounds is the abstraction of a hydrogen atom from another molecule:

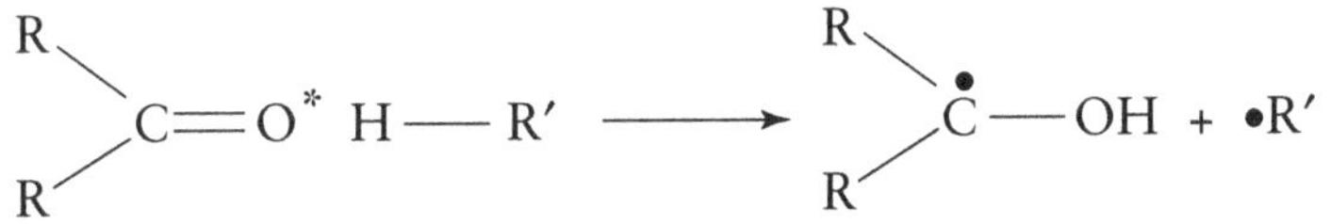

In each case the weakest C-H bond in the R'-H molecule is broken. A consequence of this reaction is that the range of suitable organic solvents for carbonyl photochemistry is somewhat limited. Cyclohexane is a relatively poor hydrogen donor although it has limited success in dissolving polar compounds, so acetonitrile and tertiary alcohols may be used, as is benzene.

The pair of radicals that is initially generated can react in a number of ways:

• The radicals may combine with each other:

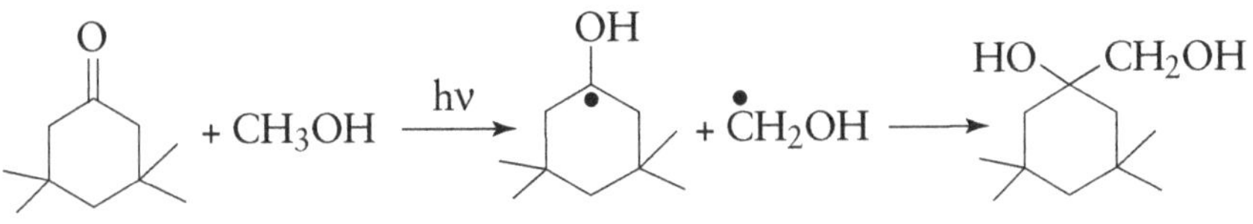

- The radicals may dimerise:

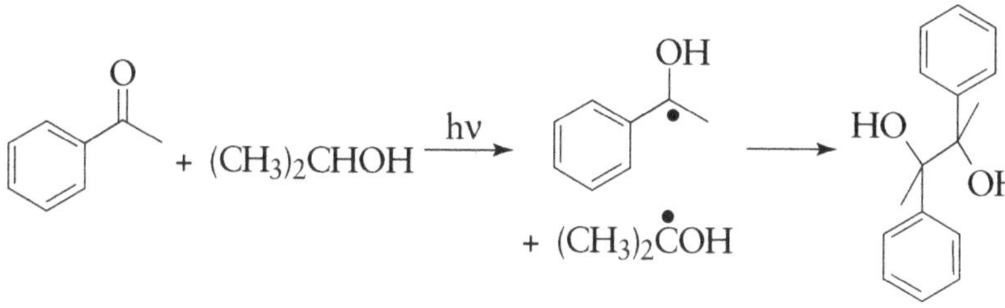

- The radical derived from the carbonyl compound may abstract a second hydrogen, either from a molecule of RH or from R•:

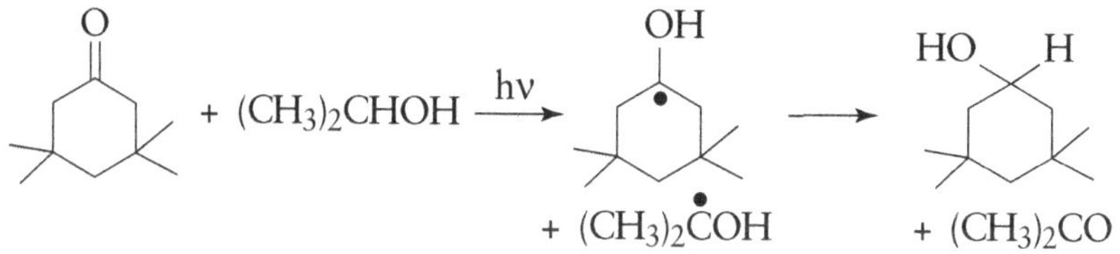

9.4 INTRAMOLECULAR HYDROGEN-ABSTRACTION REACTIONS

Hydrogen abstraction can also occur from within the excited carbonyl molecule, with the reaction occurring preferentially to give a **1,4-biradical** where the unpaired electrons are four atoms apart:

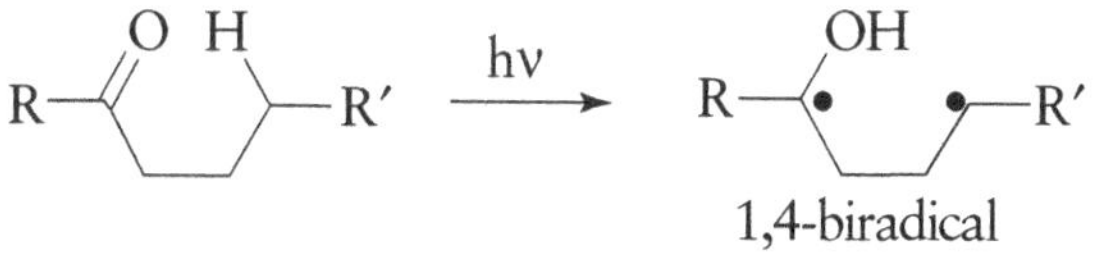

1,4-biradical

The 1,4 biradical can undergo cyclisation (**Yang cyclisation**) to give a four-membered cyclic alcohol:

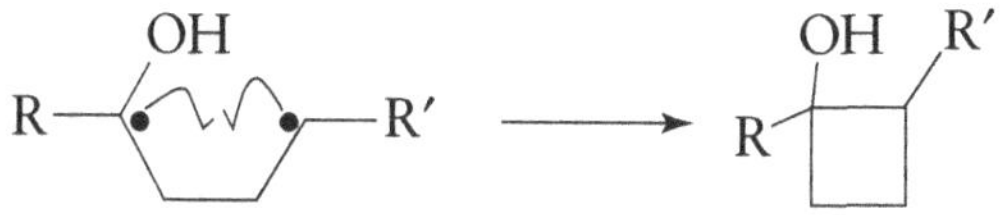

Alternatively, bond cleavage of the 2-3 C–C bond between the radical centres can occur to form an alkene and an enol. The enol rapidly isomerises to a ketone with fewer carbon atoms than the parent ketone. The reaction is regarded as a photoelimination reaction, commonly referred to as the **Norrish type 2 reaction.**

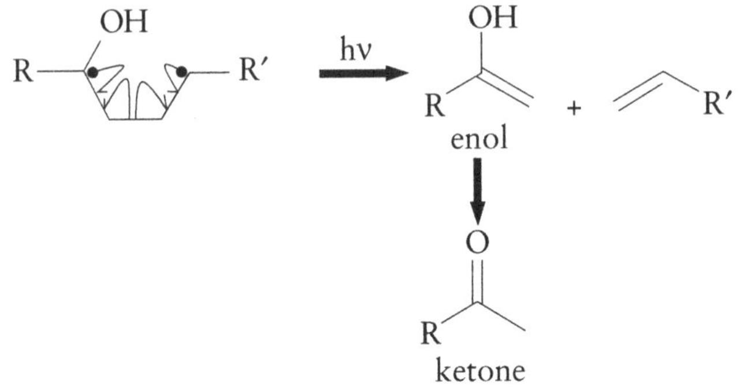

9.5 PHOTOCYCLOADDITION REACTIONS

We now come to the third principal reaction of the ketone (n,π^*) excited state: addition to alkenes.

On irradiation with alkenes, carbonyl compounds undergo photocycloaddition to give oxetanes in the **Paterno–Büchi reaction:**

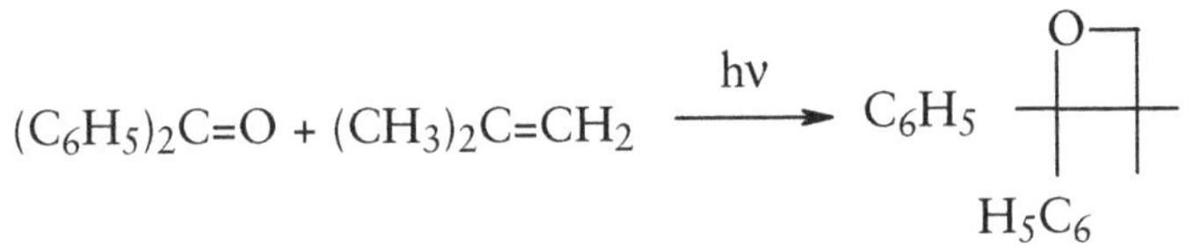

The first step of the reaction involves the (n, π^*) excited state of the carbonyl compound reacting with the ground-state alkene. For aromatic ketones, rapid intersystem crossing from the excited singlet state to the excited triplet state occurs, forming initially a 1,4-biradical and then the oxetane:

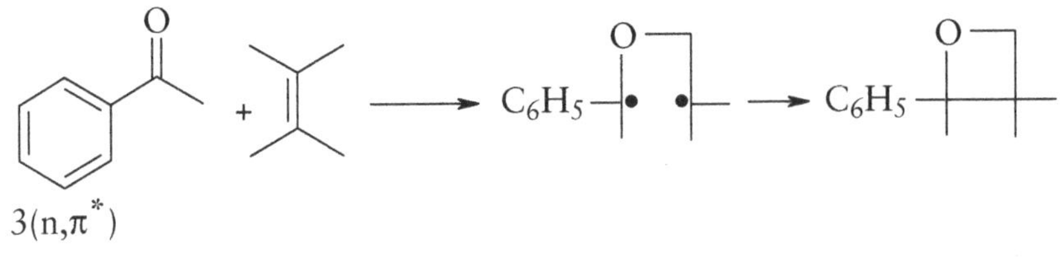

The reaction between an asymmetrical alkene and an aromatic ketone gives two different orientations of cycloaddition through two different 1,4-biradicals. The route through the more stable biradical produces the major product:

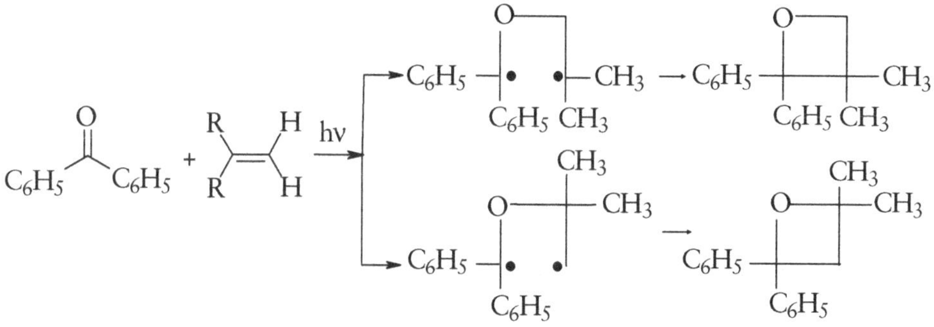

If the alkene can exist as cis and trans isomers then we need to be aware of the stereospecificity of the reaction. If the reaction involves an excited triplet state then the biradical formed will be able to undergo bond rotation in the lifetime of the excited state. The reaction is, therefore, nonstereospecific, forming a mixture of oxetane isomers from either alkene isomer:

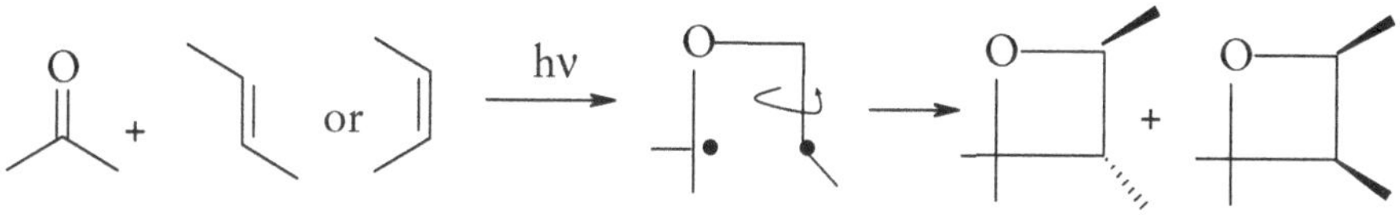

For aliphatic ketones, although intersystem crossing is fast, the singlet (n,π^*) state will react by a stereospecific process with alkenes that have electron-withdrawing groups in their molecules:

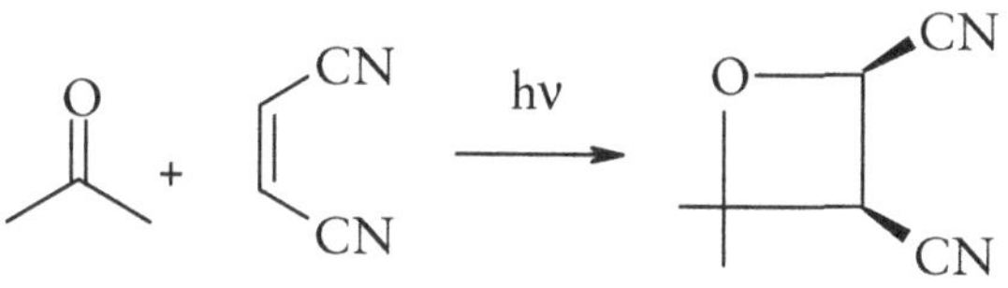

9.6   THE ROLE OF CARBONYL COMPOUNDS IN POLYMER CHEMISTRY

The photochemistry of carbonyl compounds plays an important part in the photochemical formation and breakdown of polymers.

9.6.1   Vinyl Polymerisation

The polymerisation of vinyl monomers can be initiated by radicals formed by α-cleavage or hydrogen abstraction of a carbonyl compound.

$$\text{Initiator} \xrightarrow{\ h\nu\ } 2X^{\cdot}$$

The radical formed then reacts with the unsaturated monomer and a sequence of reactions is set up (propagation) in which the chain length of the polymer grows (Scheme 9.3).

The chain-propagation sequence can be stopped by a termination step. In radical systems, radical combination and disproportionation are important mechanisms by which this occurs.

9.6.2   Photochemical Cross-linking of Polymers

It is possible to improve the mechanical strength and weather-resistance of polymers by inducing a certain degree of cross-linking.

Cross-linking in polyethene may be induced in the presence of a sensitiser such as benzophenone. The mechanism is outlined in Scheme 9.4, where the symbol P_n-H represents any C-H bond in polyethene.

$$X\bullet \ + \ H_2C{=}CHR \longrightarrow XH_2C{-}\overset{\bullet}{C}HR$$

$$XH_2C{-}\overset{\bullet}{C}HR \ + \ H_2C{=}CHR \longrightarrow XH_2C{-}CHR{-}H_2C{-}\overset{\bullet}{C}HR$$

$$XH_2C{-}CHR{-}H_2C{-}\overset{\bullet}{C}HR \ + \ H_2C{=}CHR \longrightarrow XH_2C{-}CHR{-}H_2C{-}CHR{-}H_2C{-}\overset{\bullet}{C}HR$$

Scheme 9.3

$$(C_6H_5)_2C{=}O \xrightarrow[\text{isc}]{h\nu} (C_6H_5)_2C{=}O\ (T_1)$$

$$(C_6H_5)_2C{=}O\ (T_1) \ + \ P_n\text{-H} \longrightarrow (C_6H_5)_2\overset{\bullet}{C}{-}OH \ + \ \overset{\bullet}{P}_n$$

$$\overset{\bullet}{P}_n \ + \ \overset{\bullet}{P}_n \longrightarrow P_n\text{-}P_n$$

Scheme 9.4

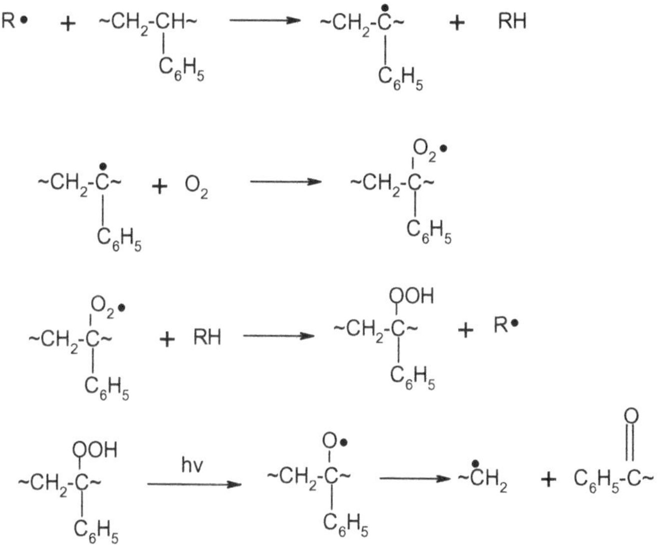

R• + ~CH$_2$-CH~ $\longrightarrow$ ~CH$_2$-Ċ~ + RH
         | |
         C$_6$H$_5$ C$_6$H$_5$

~CH$_2$-Ċ~ + O$_2$ $\longrightarrow$ ~CH$_2$-C~
      | |
      C$_6$H$_5$ C$_6$H$_5$
                          O$_2$•

~CH$_2$-C~ + RH $\longrightarrow$ ~CH$_2$-C~ + R•
    O$_2$•   OOH
    | |
    C$_6$H$_5$ C$_6$H$_5$

~CH$_2$-C~ $\xrightarrow{h\nu}$ ~CH$_2$-C~ $\longrightarrow$ ~ĊH$_2$ + C$_6$H$_5$-C~
    OOH   O•   ‖
    | | O
    C$_6$H$_5$ C$_6$H$_5$

Scheme 9.5

H$_2$C=CH-C$_6$H$_5$ + H$_2$C=CHC$_6$H$_5$ $\longrightarrow$ ~CH$_2$-CH-CH$_2$-CH~
   ‖ |
   O   C$_6$H$_5$
                                                 C=O
                                                  |
                                                 C$_6$H$_5$

copolymer

~CH$_2$-CH-CH$_2$-CH~ $\xrightarrow{h\nu}$ ~CH$_2$-CH-CH$_2$-Ċ~   (1,4- hydrogen abstraction)
      | |
      C$_6$H$_5$ C$_6$H$_5$
   C=O   •C-OH
    | |
   C$_6$H$_5$ C$_6$H$_5$

~CH$_2$-CH-CH$_2$-Ċ~ $\longrightarrow$ ~CH$_2$-CH + ~C=CH$_2$ (Norrish type 2 reaction)
      | | |
      C$_6$H$_5$ C   C$_6$H$_5$
   •C-OH   HO   C$_6$H$_5$
      |
      C$_6$H$_5$

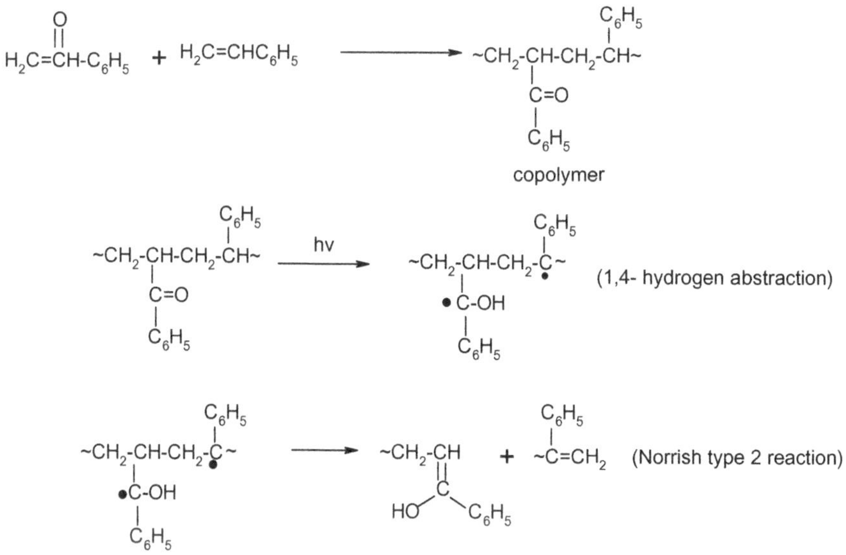

Scheme 9.6

9.6.3   Photodegradation of Polymers

Most organic polymers undergo degradation when exposed to ultraviolet light. Radicals generated photochemically from carbonyl impurities in the polymer start the breakdown, followed by reaction of the radicals with oxygen from the atmosphere.

Consider the photodegradation of polystyrene in sunlight (Scheme 9.5). Polymer photodegradation can be slowed down by adding photostabilisers, which may be:

- Quenchers to inhibit the formation of radicals.
- Scavengers to react with the radicals.
- Absorbers or light scatterers to reduce light absorption by the impurities in the polymer.

Polymers which undergo accelerated photodegradation in the environment can be made by making a copolymer of the monomer with an unsaturated ketone. The principle of the method is shown in Scheme 9.6, using styrene as the monomer.

10

Investigating Some Aspects of Photochemical Reaction Mechanisms

AIMS AND OBJECTIVES

After you have completed your study of all the components of Chapter 10, you should be able to:

- Show how some of the special features of photochemical reaction mechanisms are investigated.
- Derive information relevant to mechanistic studies of singlet states (from UV-visible absorption spectra and from fluorescence data) and triplet states (from phosphorescence data).
- Recognise features which relate to the elucidation of the nature, energy and lifetime of excited-state species.
- Describe the application of flash-photolysis techniques to the mechanistic study of a number of chemical reactions.
- Explain the usefulness of quenching and sensitisation studies to an understanding of mechanistic problems and in providing energy and lifetime data.
- Explain the usefulness of matrix isolation in mechanistic studies and give examples of intermediates which can be characterised using this technique.

Principles and Applications of Photochemistry   Brian Wardle
© 2009 John Wiley & Sons, Ltd

10.1 INTRODUCTION

When considering the manner in which photochemical reactions occur in Chapter 7, the overall reactions were considered as the sum of a number of elementary steps. It is assumed that absorption of a photon by a reactant molecule, R, produces an electronically-excited-state species, R^*, which may then react via reactive ground-state intermediate(s), I, to eventually form the product(s), P:

$$R \xrightarrow{\text{hv}} R^* \rightarrow I \rightarrow P$$

An alternative mechanism occurs for concerted photoreactions in which the bond-breaking and bond-forming processes occur together. Such reactions proceed via a single transition state:

$$R \xrightarrow{\text{hv}} R^* \rightarrow P$$

In order to investigate the elementary steps of photochemical reactions and establish the likely mechanisms involved, it is necessary to answer a number of questions:

- What is the spin multiplicity of the initial excited state?
- What is the electronic configuration – (π,π^*), (n,π^*), etc. – of the initial excited state?
- What are the energies of the excited states?
- What is the lifetime of the reactive excited state?
- What is the rate constant for the primary photochemical step $(R^* \rightarrow I)$?
- What reactive intermediate(s) occur along the reaction pathway?

By the end of this chapter you should know all the experimental techniques employed to answer these questions.

10.2 INFORMATION FROM ELECTRONIC SPECTRA

The reactive excited state in a photochemical reaction is usually either the S_1 state or the T_1 state of the reactant molecule. These states can be characterised by reference to the absorption and emission spectra of the reactant.

- The longest wavelength band in the electronic absorption spectrum of the reactant corresponds to formation of the lowest excited singlet state, since it arises from a $S_0 \rightarrow S_1$ transition.
- The nature of the $S_0 \rightarrow S_1$ transition – $(n \rightarrow \pi^*)$, $(\pi \rightarrow \pi^*)$, etc. – can be determined from the magnitude of the molar absorption coefficient or from the effect of solvent polarity on the absorption maximum (see Section 2.4).
- Absorption spectra can provide information relating to the energy of an excited singlet state. This corresponds to the lowest 0–0 vibrational transition in the electronic absorption spectrum. When the vibrational fine structure is evident, the energy of the excited singlet state is readily determined, but when the 0–0 band cannot be located, the value can be taken from the region of overlap of the absorption and fluorescence spectra.

Generally speaking, luminescence spectra (fluorescence and phosphorescence) provide more information about excited states than do absorption spectra. This is because luminescence measurements are much more sensitive than absorption measurements, and the two types of emission can be studied separately due to their widely differing lifetimes.

Excited-state lifetimes can be measured directly by monitoring the decay of luminescence, but impurities present affect both the lifetime and the luminescence spectrum. Also, because low temperatures are necessary for phosphorescence studies, the excited-state properties determined may differ from those at room temperature.

- Using the fluorescence spectrum, the energy of the emitting singlet state can be determined from the wavelength of the shortest-wavelength (longest-wavenumber) band or from the region of overlap of the fluorescence and absorption spectra, as noted above. The lifetime of the emitting singlet state is determined by monitoring the decay of fluorescence (by time-correlated single-photon counting, see Section 3.3). The energy of the emitting triplet state is determined from the phosphorescence spectrum and its lifetime is measured by monitoring the phosphorescence decay (see Section 3.3).
- The excited-state configuration – $(n \rightarrow \pi^*)$, $(\pi \rightarrow \pi^*)$, etc. – may sometimes be determined from luminescence properties. The typical radiative lifetimes of $^1(n \rightarrow \pi^*)$ singlet states are 10^{-6}–10^{-3} s and those of $^1(\pi \rightarrow \pi^*)$ states are 10^{-9}–10^{-6} s, whereas typical triplet radiative lifetimes are 10^{-4}–10^{-2} s for $^3(n \rightarrow \pi^*)$ states and 1–10^2 s for $^3(\pi \rightarrow \pi^*)$ states.

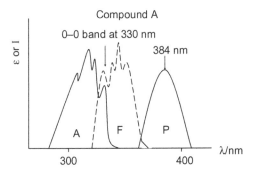

Figure 10.1 The absorption (A), fluorescence (F) and phosphorescence (P) spectra of compound A

- The ΔE_{S-T} energy gap between the singlet and triplet states is much smaller for $(n \rightarrow \pi^*)$ states ($<60\,kJ\,mol^{-1}$) than for $(\pi \rightarrow \pi^*)$ states ($>60\,kJ\,mol^{-1}$).

Figure 10.1 shows the absorption and luminescence spectra of a compound A.

The energy of the first excited singlet state can be determined from the wavelength at which the first vibrational band in the absorption spectrum coincides with the vibrational band in the fluorescence spectrum:

$$E(S_1) = N_A hc/\lambda = 1.2 \times 10^5/330 = 364\,kJ\,mol^{-1}$$

There is no vibrational information in the phosphorescence spectrum. It is assumed that 0–0 is close to the maximum at 384 nm:

$$E(T_1) = 1.2 \times 10^5/384 = 313\,kJ\,mol^{-1}$$

Although spectroscopic studies provide useful information about excited states, they do not give any information regarding which of these states (if any) are the chemically-reactive states. Energy-transfer studies involving quenching and sensitisation are very often found to be helpful in such cases.

10.3   TRIPLET-QUENCHING STUDIES

When a photochemical reaction is quenched by a compound that is known to be a triplet quencher, the chemically-active state is very likely to be an excited triplet state rather than an excited singlet state. Similarly, if the quencher also reduces the phosphorescence intensity then the reactive triplet is almost certain to be the same as the luminescent triplet.

Triplet quenching studies can provide information about:

- The involvement of a triplet state in a photochemical reaction.
- The lifetime of the triplet state involved.
- The energy of the triplet state involved.

The inhibition of product formation by a triplet quencher readily allows a triplet mechanism to be identified where the absence of inhibition by a triplet quencher indicates a singlet mechanism.

Suitable triplet quenchers should have the following properties:

- The triplet energy of the quencher must be lower than that of the T_1 state of the reactant.
- The singlet energy of the quencher must be higher than that of the S_1 state of the reactant in order to avoid singlet energy quenching by collisional energy transfer.
- The quencher should absorb much less strongly than the reactant at the wavelength of irradiation.

Conjugated dienes, such as penta-1,3-diene $H_2C=CH-CH=CH_3$, are excellent triplet quenchers.

For example, penta-1,3-diene has singlet energy $E_S = 430\,\mathrm{kJ\,mol^{-1}}$ and triplet energy $E_T = 247\,\mathrm{kJ\,mol^{-1}}$ and it will therefore act as a triplet quencher for N-methylphthalimide (1) (Figure 10.2) with $E_S = 335\,\mathrm{kJ\,mol^{-1}}$ and $E_T = 288\,\mathrm{kJ\,mol^{-1}}$.

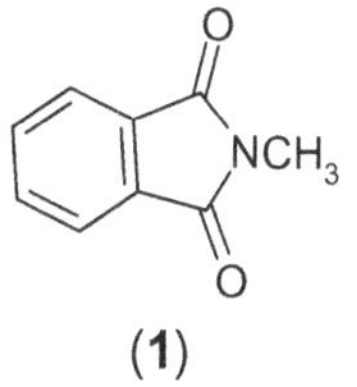

(1)

$E_T(\text{penta-1,3-diene}) < E_T(1)$

$E_S(\text{penta-1,3-diene}) > E_S(1)$

Figure 10.2   N-methylphthalimide

In addition, penta-1,3-diene and (1) will almost certainly have different absorption characteristics, due to the different chromophoric groups in the two compounds.

In Chapter 6 the Stern–Volmer equation was introduced in relation to fluorescence quenching. The corresponding equation relating to triplet-state quenching is:

$$\phi/^Q\phi = 1 + k_Q{}^3\tau[Q]$$

where ϕ and $^Q\phi$ are the quantum yields of product formation without and with the quencher, respectively; k_Q is the rate constant for quenching ; [Q] is the concentration of quencher; $^3\tau$ is the triplet lifetime in the absence of the quencher. The ratio $\phi/^Q\phi$ is equal to the ratio of the amounts of product formed in the absence and in the presence of the quencher under the same irradiation conditions. In order to ensure that each solution receives an equal intensity of radiation, an annular reaction tube is employed, where all tubes are equidistant from the light source and a 'carousel' rotates about the central axis, ensuring that any differences in light intensity are averaged out.

The Stern–Volmer equation has a linear form and the quantity $k_Q{}^3\tau$ is obtained as the slope of the plot of $\phi/^Q\phi$ against [Q] for different quencher concentrations.

Because quenching involves the collision of two molecules, its rate is effectively controlled by diffusion processes. The values of k_Q for various solvents and temperatures (commonly of the order of $k_Q = 10^9$–$10^{10}\,mol^{-1}\,dm^3\,s^{-1}$) are well documented; hence a value for the triplet lifetime may be obtained.

In some photochemical reactions both S_1 and T_1 take part in the reaction. For example, both the S_1 and T_1 states of aliphatic ketones, such as hexan-2-one, take part in Norrish type 2 reactions:

Stern–Volmer analysis for the quenching of the Norrish type 2 reaction of hexan-2-one by penta-1,3-diene results in a nonlinear plot

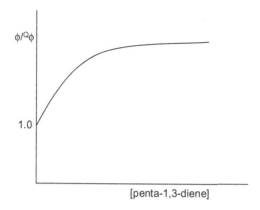

Figure 10.3 Stern–Volmer plot for the quenching of the Norrish type 2 photoreaction of hexan-2-one by penta-1,3-diene

(Figure 10.3). Preferential quenching of the T_1 state of the hexan-2-one is brought about by use of the penta-1,3-diene and it can be assumed that at high concentrations of the diene all the T_1 state and none of the S_1 state will be quenched. For this situation, $\phi/^Q\phi$ will reach a limiting value at the plateau of Figure 10.3, corresponding to the reaction of S_1, and will not vary with Q at high values of [Q].

It is possible to obtain a Stern–Volmer plot specifically for the triplet reaction from the equation:

$$\phi^T/^Q\phi^T = \phi/^Q\phi - (\phi/^Q\phi)_{\text{plateau}}$$
$$T_1 = T_1 + S_1 - S_1$$

Thus a plot of $\phi^T/^Q\phi^T$ against [Q] for low values of [Q] should be a linear plot with intercept 1 and slope for the triplet reaction, from which the rate constant, $k_R = 1/k_Q^3\tau$, may be determined.

If several different triplet quenchers of different triplet energy are used and the efficiency of the photochemical reaction is determined, a value for the energy of the reactive T_1 state of the reactant can be obtained. By considering the relative efficiency of quenching for each quencher and knowing its triplet energy, it will be found that the efficiency of quenching will be markedly reduced at a certain quencher value (and above). The energy of the reactive triplet state of the reactant will then be close to this value.

10.4   SENSITISATION

When energy transfer is used to form an excited state rather than to quench it, the process is called sensitisation.

A T_1 state has a relatively long lifetime compared to the corresponding S_1 state and so it can undergo efficient quenching, transferring energy to the ground-state reactant molecule.

$$^3D^* + {}^1R \rightarrow {}^1D + {}^3R^*$$

In order for efficient sensitisation to occur, the energy of the T_1 state of the sensitiser must be greater than that of the reactant and the energy of the S_1 state of the sensitiser must be less than that of the reactant (see Section 6.7). In addition, the triplet sensitiser should be strongly absorbing at the wavelength of irradiation and undergo efficient S $\rightsquigarrow$ T intersystem crossing. Aromatic ketones such as acetophenone, $C_6H_5COCH_3$ and benzophenone, $(C_6H_5)_2CO$, are excellent triplet sensitisers.

Triplet-sensitisation studies can provide information regarding:

- The involvement of a triplet state in a photochemical reaction.
- The energy of the triplet state.

If the products formed from the direct and sensitised reactions differ then the triplet state formed on sensitisation is not the reactive excited state formed by direct irradiation. This situation is illustrated in Sections 8.2 and 8.3, where direct and photosensitised cis–trans isomerisation of alkenes was considered.

If the same products are formed from the direct and sensitised reactions then the triplet state formed on sensitisation may be the state responsible for reaction on direct irradiation. It is important to understand that it is possible for the triplet state to give the same products as the reactive singlet state.

Energy-transfer experiments involving sensitisation and quenching allow us to selectively populate and depopulate excited states. This affords us a powerful tool whereby the particular excited state responsible for a photochemical reaction may be identified. The excited state responsible for the photosolvation reaction of $Cr(CN)_6^{3-}$ in dimethylformamide (DMF) solution at room temperature has been identified by carrying out sensitisation and quenching experiments. Direct excitation of the complex to one of its excited quartet states results in a photosolvation reaction and phosphorescence:

$$Cr(CN)_6^{3-} + DMF + hv \rightarrow Cr(CN)_5DMF^{2-} + CN^-$$

$$Q_1 \rightsquigarrow D_1 \rightarrow D_0 + hv$$

Sensitisation by a high-energy donor, such as triplet xanthone, can populate both the Q_1 and the D_1 states by energy transfer (Figure 10.4). This results in both the photosolvation reaction and phosphorescence emission.

Sensitisation with lower-energy donors, such as triplet $Ru(bpy)_3^{2+}$, on the other hand, means that energy transfer is only possible to the D_1 state (Figure 10.4). This results in phosphorescence emission, but the photosolvation reaction does not occur. This shows that the Q_1 state must be responsible for the photosolvation reaction.

Quenching experiments show that the phosphorescence emission is strongly quenched by oxygen, whereas the presence of oxygen has no

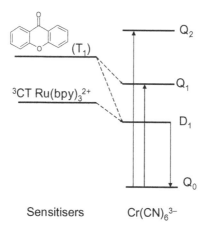

Figure 10.4 Determining the reactive excited state of $Cr(CN)_6^{3+}$ by sensitisation Adapted from F. Scandola and V. Balzani, 'Energy-Transfer Processes of Excited States of Coordination Compounds', *Journal of Chemical Education*, Volume 60 (10), 1983. © American Chemical Society

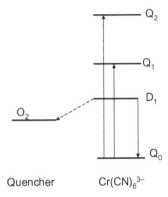

Quencher $Cr(CN)_6^{3-}$

Figure 10.5 Quenching by oxygen affects the phosphorescence emission arising from D_1 but has no effect on the photosolvation reaction that occurs from Q_1 Adapted from F. Scandola and V. Balzani, 'Energy-Transfer Processes of Excited States of Coordination Compounds', *Journal of Chemical Education*, Volume 60 (10), 1983. © American Chemical Society

effect on the photosolvation reaction. This situation is illustrated in Figure 10.5.

10.5 FLASH PHOTOLYSIS STUDIES

Methods used for the direct characterisation of reactive intermediates must have a timescale shorter than or comparable to the lifetime of the reactive intermediate (Table 10.1). Because of the very short lifetimes of these transient intermediates, this can be brought about by:

- Making the technique sufficiently fast that it is able to detect and measure some feature of R^* or I during its lifetime.
- Isolating the intermediate under conditions whereby its lifetime is long enough for it to be observed (such as matrix isolation, see Section 10.6).

The technique of **flash photolysis** (Figure 10.6) employed by Norrish and Porter in 1949 revolutionised the study of short-lived transient species as it proved capable of generating and analysing chemical species with lifetimes shorter than a few milliseconds.

The basis of flash photolysis is to irradiate the system with a very short, intense pulse of light and then, as soon as the pulse is over, to monitor the changes in the system with time by some spectroscopic

Table 10.1 Monitoring ultra-fast events in photochemistry

Timescale	Examples
millisecond (1 ms = 10^{-3} s)	Lifetimes of radicals and excited triplets
microsecond (1 μs = 10^{-6} s)	
nanosecond (1 ns = 10^{-9} s)	Lifetime of excited singlets
picosecond (1 ps = 10^{-12} s)	Energy-transfer and electron-transfer processes
femtosecond (1 fs = 10^{-15} s)	Primary events in photosynthesis

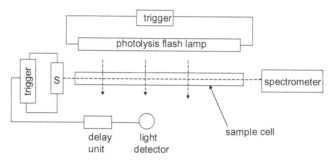

S is the spectroscopic flash lamp

Figure 10.6 The original (Norrish and Porter) flash photolysis apparatus

method. The high-intensity photolysis pulse produces a large number of photons, which in turn produce a large electronically-excited-state population, the time evolution of which can be monitored by absorption or emission spectroscopy. The monitoring device must have a time resolution fast enough to observe the transient state before it decays. The development of millisecond and microsecond flash photolysis allowed detailed study of photochemical reaction kinetics and mechanisms relating to a number of free-radical species and electronically-excited states.

With the invention of the laser in 1960 and the subsequent development of pulsed lasers using Q-switching (Chapter 1), monochromatic and highly-collimated light sources became available with pulse durations in the nanosecond timescale. These Q-switched pulsed lasers allow the study of photo-induced processes that occur some 10^3 times faster than events measured by flash lamp-based flash photolysis.

By the late 1960s the development of mode locking (Chapter 1) allowed the study of picosecond laser techniques. Excited-state processes carried out in the picosecond domain allow such processes as intersystem crossing, energy transfer, electron transfer and many photoinduced unimolecular reactions to be investigated.

A laser excitation source that is employed in many laboratories for nanosecond flash photolysis studies is the Q-switched Nd-YAG laser. This device typically has a pulse width of 6 ns and lases at its fundamental wavelength of 1064 nm. The 1064 nm line can be frequency-doubled to 532 nm, tripled to 3555 nm or quadrupled to 266 nm to give excitation wavelengths in the visible and ultra violet. Frequency doubling is achieved by passing the beam through a crystal of potassium dihydrogen phosphate (KDP).

According to the Beer–Lambert law (Chapter 2), when a beam of light of wavelength λ is incident on an absorber, the logarithm of the ratio of incident light intensity, I_0, to that of the transmitted light intensity, I, is termed the absorbance, A_λ, of the sample. A_λ is related to the molar concentration, c, of the absorbing species according to:

$$A_\lambda = \log(I_0/I) = \varepsilon c \ell$$

where ε is the molar absorption coefficient at wavelength λ and ℓ is the optical path length of the sample.

Now the concentration of a transient absorbing species varies with time, c(t), and so A is a function of two variables, namely time and wavelength:

$$A(\lambda, t) = \log(I_0/I) = \varepsilon c(t) \ell$$

The wavelength-dependent data provide spectral information, useful for assignment or structural purposes, while the time-dependent data lead to information relating to the kinetics of the processes that occur in the sample.

Thus, two distinct types of experiment have been employed to generate $A(\lambda, t)$ data for transient species:

- The **spectroscopic technique** involves measurement of a complete absorption spectrum at a specific time after excitation. The spectrally-resolved absorption kinetics can be obtained by stepping the delay through a predetermined time range and digitally storing the data. Details of the transient absorption kinetics are obtained by measuring the point on each curve corresponding to the light absorbed at a particular wavelength from a large number of such experiments.
- The **kinetic technique** measures the decay of the transient species as a function of time at a single wavelength. The transient decay

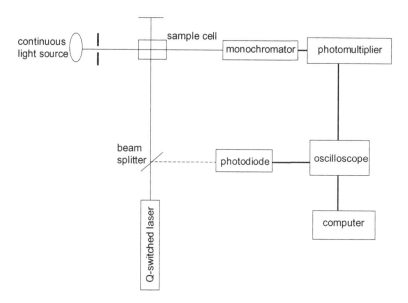

Figure 10.7 Nanosecond kinetic flash photolysis apparatus

is digitally stored and the experiment repeated over a range of wavelengths. By measuring the point on each curve corresponding to the light absorbed at a particular delay time from a large number of such experiments it is possible to generate the absorption spectrum of the transient species at different times after the initial excitation.

Figure 10.7 shows a nanosecond kinetic flash photolysis apparatus. The absorbance of the sample is monitored, using a photomultiplier, by the change in the transmittance of the sample to the xenon arc lamp continuous light source.

The electronic devices used in nanosecond flash photolysis are at the limit of their time responses to the signals they receive. In order to investigate reactions occurring in the sub-nanosecond timescale it is necessary to overcome this problem.

The **pump-probe** method provides the solution to this nanosecond barrier. Here, two light pulses are generated: one to excite the sample (prepare the excited state) and one to probe the system at a given time after excitation (Figure 10.8).

The mode-locked laser output is split into two parts by use of a partially-reflecting mirror or beam splitter. The two pulses leave the

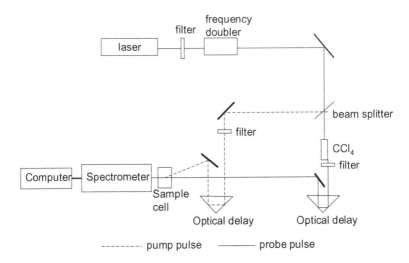

Figure 10.8 Picosecond kinetic flash photolysis apparatus

beam splitter at the same time but then travel along different pathways. The path length each pulse travels is determined by a series of mirrors, which are arranged so that the pulses arrive at the sample at different times. In a time-resolved laser experiment the pulse which initiates the photochemical change is called the **pump pulse** and the pulse used to record the changes in the sample brought about by the pump pulse is called the **probe pulse**.

A broad band emission for the probe pulse is obtained by focussing part of the monochromatic laser pulse into a cell containing a suitable liquid, such as water or tetrachloromethane. The liquid converts the laser beam into a polychromatic pulse covering a wide range of wavelengths through much of the UV, visible and IR regions.

Events that have been studied in the picosecond timescale include isomerisation, internal conversion, energy transfer and electron transfer.

10.5.1   An Aside: Some Basic Ideas on Reaction Kinetics

Most photochemical reactions are composite reactions; that is, they can be regarded as being made up of several elementary steps, constituting the mechanism, each of which involves a simple number of species. The

molecularity of an elementary step is the total number of species that react together.

A valid mechanism for a reaction should be based on sound experimental evidence acquired through a study of the way in which the various intermediates involved affect the kinetics of the reaction. Experimentally-determined **rate laws** provide us with a means of classifying reactions according to their kinetics. For example, for a reaction $X + Y \rightarrow Z$, if the experimentally-determined rate law allows us to determine the **order** of the reaction:

$$Rate = k[X][Y]^2$$

The reaction is first-order with respect to X, second-order with respect to Y and third-order overall; k is the **rate constant**. Note that the order of a reaction does not necessarily have to have integral values.

Simplification of a rate law can be achieved by use of the **isolation method**, in which all reactants except one are present in large excess. If a reactant is present in large excess, its concentration may be considered to be constant throughout the course of the reaction. For example, for the rate law:

$$Rate = k[X][Y]^2$$

If Y is in large excess, we can rewrite the rate law as:

$$Rate = k'[X]$$

where $k' = k[Y]_0^2$ and $[Y]_0$ is the initial concentration of Y.

In this case, the new rate law has the form of a first-order rate law and is classified as a **pseudo first-order rate law**.

Simple **integrated rate laws** for single reactants allow us to express the rate of reaction as a function of time. These are summarised in Table 10.2.

10.5.2 Flash Photolysis Studies in Bimolecular Electron-transfer Processes

By using pulsed laser sources and fast measuring devices, direct observation of redox products in flash photolysis experiments provides evidence regarding oxidative and reductive quenching mechanisms in $^*Ru(bpy)_3^{2+}$.

Table 10.2 Integrated rate laws for first- and second-order reactions

Order	Reaction and Rate Law	Integrated Rate Law	Plot
1	$X \rightarrow Z$ Rate $= -k[X]$	$\ln[X] - \ln[X]_0 = -kt$	$\ln[X]$ vs t gives slope $= -k$
2	$2X \rightarrow Z$ Rate $= -k[X]^2$	$1/[X] - 1/[X]_0 = kt$	$1/[X]$ vs t gives slope $= k$

Table 10.3 Absorption and emission features of tris-2-2'-bipyridylruthenium species

Species	Spectral Features
$Ru(bpy)_3^+$	Intense absorption band ~500 nm
$Ru(bpy)_3^{2+}$	Strong MLCT absorption ~430 nm
$Ru(bpy)_3^{3+}$	Weakly absorbing
$^*Ru(bpy)_3^{2+}$	Luminescence band ~615 nm

Determination of changes in the absorbance (ΔA) of a solution of the complex containing a quencher can be investigated when a pulse of laser light excites $Ru(bpy)_3^{2+}$. The principal spectral characteristics of the complex and its derived transient species are given in Table 10.3.

The difference spectrum of a solution of $Ru(bpy)_3^{2+}$ and Eu^{2+} obtained by a pulse of several ns from a frequency-doubled Nd-YAG laser at 530 nm is shown in Figure 10.9.

The difference spectrum shows:

- Net loss of absorbance (bleaching) in the region around 430 nm due to depletion of the $Ru(bpy)_3^{2+}$.
- A net gain in absorbance at 490–510 nm due to formation of $Ru(bpy)_3^+$ (both Eu^{2+} and Eu^{3+} show no absorption in this region).

These features are consistent with the reductive quenching mechanism of $^*Ru(bpy)_3^{2+}$:

$$^*Ru(bpy)_3^{2+} + Eu^{2+} \rightarrow Ru(bpy)_3^+ + Eu^{2+}$$

Similarly, kinetic spectroscopy studies (Figure 10.10) show the growth of absorption of $Ru(bpy)_3^+$ at 500 nm occurs at the same rate as the decay of luminescence of $^*Ru(bpy)_3^{2+}$.

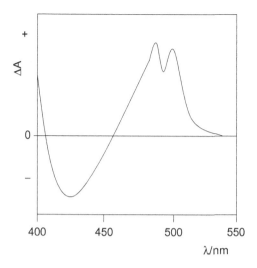

Figure 10.9 Difference absorption spectrum for $(Ru(bpy)_3^+ - Ru(bpy)_3^+)$ produced by pulsed excitation of a solution of $Ru(bpy)_3^{2+}$ containing Eu^{2+}
Reprinted from R.J. Watts, 'Ruthenium Polypyridyls', *Journal of Chemical Education*, Volume 60 (10), 1983. © American Chemical Society

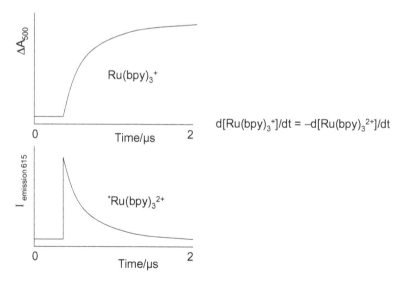

Figure 10.10 Flash photolysis studies of $Ru(bpy)_3^{2+}$ solution containing Eu^{2+}

The kinetics of the quenching of $*Ru(bpy)_3^{2+}$ by the methyl viologen species, MV^{2+}, is monitored by measuring the absorbance of the resulting strongly-absorbing blue MV^+ at 600 nm, rather than the

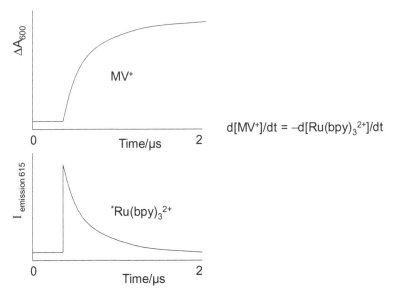

$$d[MV^+]/dt = -d[Ru(bpy)_3^{2+}]/dt$$

Figure 10.11   Flash photolysis studies of $Ru(bpy)_3^{2+}$ solution containing MV^{2+}

weakly-absorbing oxidation product. The rate of growth of the MV^+ absorption is equal to the rate of luminescence decay of $Ru(bpy)_3^{2+}$ (Figure 10.11).

These features are consistent with the oxidative quenching mechanism of $^*Ru(bpy)_3^{2+}$:

$$^*Ru(bpy)_3^{2+} + MV^{2+} \rightarrow Ru(bpy)_3^{3+} + MV^+$$

10.5.3   Photochemistry of Substituted Benzoquinones in Ethanol/Water

A further application of flash photolysis is illustrated by the reactions of a quinone (Q). The kinetic scheme in ethanol (RH) and water is made up of a number of elementary steps, which are given in Figure 10.12.

Spectroscopic flash photolysis allows the absorption wavelengths for the transient species to be found. The triplet state, $^3Q^*$, is found to have an absorption at 490 nm, and the quinone radical, QH•, has an absorption at 410 nm.

Using kinetic flash photolysis, the decay of the transient species can be determined as a function of time at the appropriate single wavelength found by the spectroscopic method.

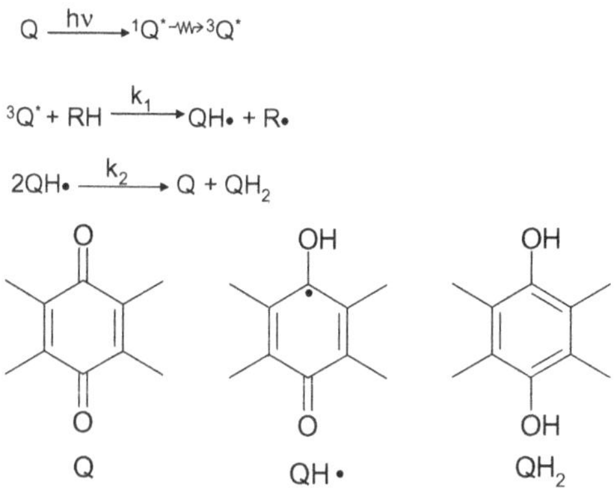

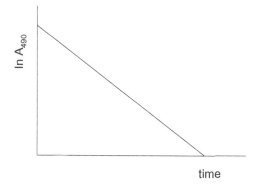

Figure 10.12   The elementary steps in the kinetic scheme of ethanol (RH) and water

Figure 10.13   Integrated rate-law plot of the decay of quinine triplet species

When RH is in large excess, the triplet state undergoes pseudo first-order reaction to form QH• and R•, so an integrated rate-law plot of $\ln[^3Q^*]$ against t gives a straight line of slope $-k_1$ (Figure 10.13).

The bimolecular reaction of the radical species, QH•, is second-order (Figure 10.14):

$$d[QH\cdot]/dt = -k_2[QH\cdot]^2$$

Substituting the absorbance $A_{410} = \varepsilon_{410}[QH\bullet]\ell$ and integrating gives:

$$1/(A_{410})_t - 1/(A_{410})_{t=0} = k_2 t/\varepsilon_{410}\ell$$

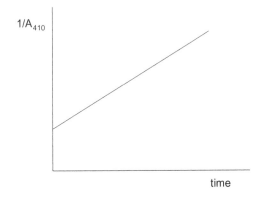

Figure 10.14   Integrated rate law plot of the second-order decay of QH• radical species

   Thus the rate constant, k_2, can be obtained from the integrated rate plot only if the excited-state absorption coefficient, ε_{410}, and the path length, ℓ, are known.

10.5.4   Time-resolved Infrared Spectroscopy

Intermediates in the photochemical reactions of metal carbonyls generally have broad featureless UV-visible absorption bands which provide little information relating to their structures. However, infrared bands associated with C-O stretching modes (1650–$2150\,cm^{-1}$) provide detailed structural information, being very intense and sharp, and the primary photochemical pathways of transition metal carbonyls in solution can be deduced from the time-resolved infrared absorption spectrum of the compound obtained at a specific time after the photolysis flash. Over many years a large amount of infrared absorption data relating to metal carbonyl species has been accumulated, allowing the characterisation of carbonyl intermediates from their absorption maxima on the time-resolved infrared spectrum.
   A kinetic trace for a particular metal carbonyl intermediate is recorded at a specific wavelength obtained from the time-resolved infrared absorption spectrum. Suitable data analysis allows determination of the kinetics of the decay of the intermediate.
   $[CpFe(CO)_2]_2$ ($Cp = \eta_5\text{-}C_5H_5$) has two photochemical pathways:

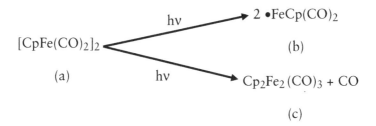

The time-resolved infrared spectrum of $[CpFe(CO)_2]_2$ in cyclohexane solution 5 μs after a UV flash shows peaks at $1938\,cm^{-1}$ and $1823\,cm^{-1}$, corresponding to (b) and (c), respectively.

Data from the kinetic trace due to radical (b) is analysed by plotting a graph of 1/(absorbance at $1938\,cm^{-1}$) against time, giving a straight line showing second-order kinetics.

$$2\cdot FeCp(CO)_2 \rightarrow [CpFe(CO)_2]_2$$

The kinetic trace due to (c) at different CO pressures shows its lifetime is dependent on the pressure of CO above the solution. Analysis of the kinetic trace data of this intermediate under constant CO pressure shows that its decay follows first-order kinetics and allows calculation of the bimolecular rate constant for its reaction with CO:

$$Cp_2Fe_2(CO)_3 + CO \rightarrow [CpFe(CO)_2]_2$$

10.5.5 Femtochemistry

By means of femtochemistry, investigation of elementary reactions on a timescale of femtoseconds ($10^{-15}\,s$) is possible. The method employs a combination of pulsed-laser and molecular-beam technologies. Investigation of a unimolecular reaction by femtosecond spectroscopy involves two ultra-fast laser pulses being passed into a beam of reactant molecules.

The first experiments involving the femtosecond timescale were carried out by Zewail on the decomposition of iodocyanide.

$$ICN + h\nu \rightarrow I + CN$$

The pump pulse (306 nm) had photons of the wavelength needed for absorption by the ground-state ICN, resulting in production of the dissociative excited state. The weaker probe pulse that arrives a few femtoseconds later has a wavelength (388 nm) capable of causing excitation of the CN dissociation product. Subsequent fluorescence from the excited CN* is then recorded, where the intensity of the fluorescence is a measure of the concentration of CN present:

$$ICN \xrightarrow[\text{pump pulse}]{\text{hv 306 nm}} I + CN$$

$$CN \xrightarrow[\text{probe pulse}]{\text{hv 388 nm}} CN^*$$

$$CN^* \xrightarrow{\hspace{3cm}} CN + hv$$
Laser-
induced
fluorescence

By varying the time delay between the pump and probe pulses, the resulting fluorescence brought about by the 388 nm pulse can be determined as shown in Figure 10.15.

Figure 10.15 shows that the fluorescence intensity increases as the delay between the pump and the probe pulses increases, until a constant level of fluorescence intensity is produced (where all the excited molecules have undergone photodissociation). The curve fitting the experimental data has the form:

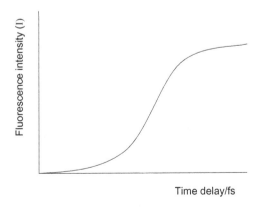

Figure 10.15 Time-resolved data plot for the laser-induced fluorescence from CN*

$$I = 1 - \exp(-t/\tau)$$

from which τ, the reaction half life, can be determined.

10.6   LOW-TEMPERATURE STUDIES

Most chemical reactions can be slowed down by lowering the temperature. With low-temperature studies it is possible to prolong the lifetimes of the reactive intermediates so that they can be characterised by normal techniques. **Matrix isolation** allows experiments to be carried out at temperatures as low as 4 K, in order to study species, such as radicals, that are produced photochemically at very low temperatures. The initial photoproduct is trapped within a rigid matrix that inhibits diffusion of the reactive species. The matrix material must be:

- Composed of molecules that are very closely packed.
- Well below its melting-point temperature.
- Transparent to the wavelengths used.
- A good solid solvent for the reactant molecules used.
- Inert toward all species present in the matrix.

The most useful matrix materials are solid argon, solid neon and solid nitrogen.

Species identified by matrix isolation include cyclobutadiene and benzyne, where products derived from these molecules are formed at higher temperatures.

FURTHER READING

G. Porter (1997) *Chemistry in Microtime: Selected Writings on Flash Photolysis, Free Radicals, and the Excited State*, Imperial College Press.

11

Semiconductor Photochemistry

AIMS AND OBJECTIVES

After you have completed your study of all the components of Chapter 11, you should be able to:

- Understand the molecular orbital structure of semiconductors.
- Explain the origin of photogenerated electrons and holes in a semiconductor irradiated with light of energy greater than the band gap.
- Understand how energy-rich electron–hole pairs can be utilised to produce electricity in photovoltaic solar cells, to drive chemical reactions where water is split into hydrogen and oxygen, to take part in photocatalytic reactions and to change the surface of the semiconductor, resulting in the phenomenon of superhydrophilicity.
- Explain how doping of an intrinsic semiconductor such as silicon leads to modification of its properties.
- Use ideas about p- and n-type silicon to explain what happens when a junction is formed between them and what effect is produced when the junction is irradiated with light of a suitable wavelength.
- Explain the photochemical principles of the dye-sensitised solar cell and understand the importance of various physicochemical parameters to the overall performance of such a cell.
- Apply simple ideas regarding energy levels to a predictive understanding of the feasibility of semiconductor-photosensitised redox reactions, including water-splitting reactions.

Principles and Applications of Photochemistry   Brian Wardle
© 2009 John Wiley & Sons, Ltd

- Understand simple energy-transfer situations, in order to explain the conditions necessary for photosensitised oxidation and reduction to occur on the surface of a TiO_2 semiconductor.
- Understand that a consideration of similar ideas about photo-excited TiO_2 leads to an explanation of a variety of effects, such as removal of pollutants in water and air, destruction of cells of pathogens and cancers, and the concept of semiconductor-photoinduced superhydrophilicity.

11.1   INTRODUCTION TO SEMICONDUCTOR PHOTOCHEMISTRY

In a solid, the molecular orbitals interact strongly to form broad energy bands. The electronic structure of semiconductors (e.g. Si, TiO_2, ZnO, $SrTiO_3$, CdS, ZnS) is characterised by a **valence band (VB)** filled with electrons and a **conduction band (CB)** in which there are no electrons. Between the bands there are no energy levels. The energy difference between the lowest energy level of the CB and the highest energy of the VB is known as the band-gap energy (Eg). Irradiation of the semiconductor with photons of energy equal to or exceeding the band-gap energy results in promotion of an electron from the VB to the CB, leaving an electron vacancy known as a **hole (h^+)** behind in the VB (Figure 11.1).

The generation of energy-rich electron–hole pairs has been utilised to:

- Produce electricity in photovoltaic solar cells (see Section 11.2).
- Drive chemical reactions in which water is split into hydrogen and oxygen (see Section 11.3).

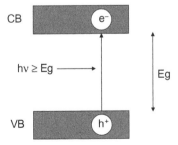

Figure 11.1   Promotion of an electron from the valence band to the conduction band on irradiation of a semiconductor. The electron vacancy left behind in the valence band is known as a hole (h^+)

- Take part in photocatalytic reactions (see Section 11.4).
- Change the surface of the semiconductor, resulting in the phenomenon of superhydrophilicity (see Section 11.5).

11.2   SOLAR-ENERGY CONVERSION BY PHOTOVOLTAIC CELLS

Electrical cells based on semiconductors that produce electricity from sunlight and deliver the electrical energy to an external load are known as **photovoltaic cells**. At present most commercial solar cells consist of silicon doped with small levels of controlled impurity elements, which increase the conductivity because either the CB is partly filled with electrons (**n-type** doping) or the VB is partly filled with holes (**p-type** doping). The electrons have, on average, a potential energy known as the **Fermi level**, which is just below that of the CB in n-type semiconductors and just above that of the VB in p-type semiconductors (Figure 11.2).

In doped silicon (an **extrinsic** semiconductor) the doping element has either three or five valence electrons (one electron less or one electron more than the four valence electrons of silicon). Substituting an arsenic or phosphorus atom (five valence electrons) for a silicon atom in a silicon crystal provides an extra loosely-bound electron that is more easily excited into the CB than in the case of the pure silicon. In such an n-type semiconductor, most of the electrical conductivity is attributed

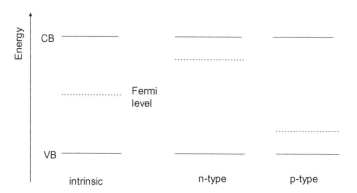

Figure 11.2   Energy diagram for various semiconductors. Shown are examples for pure (intrinsic) and doped (extrinsic) n- and p-types. In each case the Fermi level is shown as a dotted line

to the CB electrons, which are called the **majority carriers**, while the holes (which only make a small contribution to the conductivity) are called the **minority carriers**. Substituting a gallium or indium atom (three valence electrons) for a silicon atom in a silicon crystal leaves a hole from which thermally-excited electrons from the VB can move, leaving behind mobile holes, which are the majority carriers. Electrons are the minority carriers and the material is called a p-type semiconductor.

Since the current flow within a semiconductor consists of the movement of light-generated electrons and holes, it is necessary to prevent these charge carriers from recombining. In order to do this, an electrical field needs to be generated within the semiconductor and this is usually brought about by creation of a p–n junction (joining together a p-doped and an n-doped sample of silicon in one uniform sample).

When n-doped and p-doped silicon are joined together into a single semiconductor crystal, the Fermi levels of the n-type and p-type regions are aligned to the same energy level and the CB and VB bend in the region of the junction, as shown in Figure 11.3. Excess electrons move from the n-type side to the p-type side, resulting in a build up of negative charge along the p-type side and a build up of positive charge along the n-type side of the interface.

When photons are absorbed by the p–n junction, it acts as a photovoltaic cell at which electrons are promoted from the VB to the CB, forming an electron–hole pair. Connecting the p- and n-type silicon via

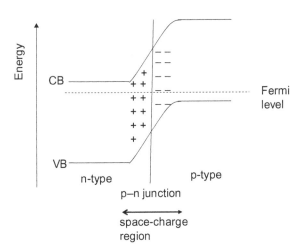

Figure 11.3 The space-charge region produced at a p–n junction

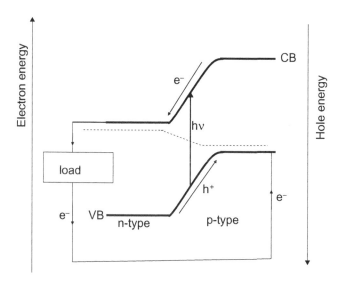

Figure 11.4   The working mechanism of a silicon p–n-junction solar cell

an external load, the electron–hole pair tends to separate rather than recombine and a resulting current is produced as the electrons pass to the n-region and the holes to the p-region.

Thus, the n-type and p-type silicon become the negative pole and the positive pole respectively of the solar cell (Figure 11.4). The change in potential energy of the electrons and holes brought about by photon absorption means that the Fermi levels in the n- and p-type silicon become separated.

While the application of photovoltaic cells has been dominated by solid-state junction devices principally made from silicon, recent work in this field offers the prospect of efficient solar-energy conversion by novel methods.

11.2.1   Dye-sensitised Photovoltaic Cells

Oxide semiconductors such as TiO_2, ZnO, SnO_2 and Nb_2O_5 have good stability but do not absorb visible light because of their relatively wide band gaps. However, by using a sensitisation process, a coloured organic dye adsorbed on to the semiconductor surface can absorb visible light and the resulting excited electrons may be injected into the CB of the semiconductor. In order to improve the light-harvesting properties and

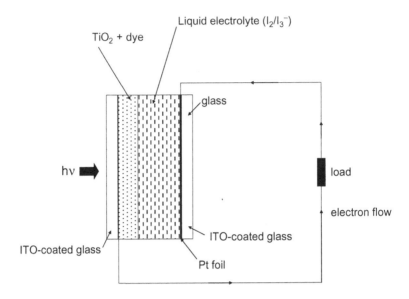

Figure 11.5 Essential features of a dye-sensitised solar cell based on sensitiser-coated TiO_2 nanoparticles

increase cell performance of these dye-sensitised solar cells (DSSCs or Grätzel cells), nanometre-sized TiO_2 particles which adsorb large amounts of dye with a broad absorption range are used. The schematic structure of a nanocrystalline DSSC is shown in Figure 11.5.

Nanoparticles of TiO_2 are deposited on to a glass support covered with a transparent conducting layer of tin-doped indium oxide (ITO). Each nanoparticle is coated with a monolayer of sensitising dye based on Ru(II). Photoexcitation of the dye results in the injection of an electron into the CB of the semiconductor.

$$Ru(II) + h\nu \rightarrow Ru(II)^*$$
$$Ru(II)^* \rightarrow Ru(III) + e^- \text{ (injected into CB of } TiO_2\text{)}$$

The electron flows through the external circuit from the ITO electrode to the Pt counter-electrode.

In order to regenerate the sensitising dye from its oxidised form, a mediator is employed. An organic solvent containing the iodide/triiodide couple is used as the mediator, by which the electron flow at the platinum counter-electrode reduces triiodide to iodide, which in turn reduces Ru(III) to Ru(II).

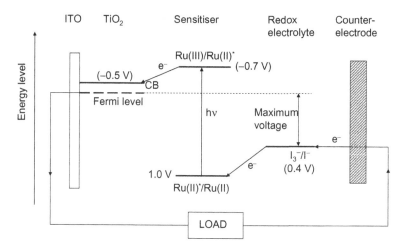

Figure 11.6 Schematic energy-level diagram relating to the principles of operation of the DSSC
Adapted from M. Gratzel, 'Solar Energy Conversion by Dye-Sensitized Photovoltaic Cells', *Inorganic Chemistry*, Volume 44 (20), 2005 © American Chemical Society

$$\text{At Pt electrode: } I_3^- + 2e^- \rightarrow 3I^-$$
$$3I^- + Ru(III) \rightarrow I_3^- + Ru(II)$$

Thus electrical power is produced without any permanent chemical change. The overall performance of the DSSC depends on the energy levels of certain of the components of the cell, namely:

- The HOMO and LUMO of the sensitising dye.
- The Fermi level of the TiO_2 electrode, which is close to the CB level.
- The redox potential of the I_3^-/I^- mediator.

The following relationships exist between the various energy levels and the performance of the DSSC:

- The HOMO–LUMO gap of the photosensitiser determines the current produced on irradiation. The smaller the size of this gap, the larger the photocurrent, due to the ability of the dye to absorb longer-wavelength regions of the solar spectrum.
- The energy level of the LUMO of the sensitiser must be higher than that of the CB of the TiO_2 to allow efficient electron injection into the CB. Similarly, the energy level of the HOMO of the sensitiser

must be lower than that of the redox couple of the mediator in order that efficient electron transfer can occur.

- The maximum voltage of the DSSC is given by the energy gap between the Fermi level of the semiconductor electrode and the redox potential of the mediator.

The DSSC differs substantially from the p–n-junction solar cell because electrons are injected from the photosensitiser into the CB of the semiconductor and no holes are formed in the VB of the semiconductor.

11.3   SEMICONDUCTORS AS SENSITISERS FOR WATER SPLITTING

The production of renewable, nonpolluting fuels by conversion of solar energy into chemical energy presents a challenge to photochemists in the twenty-first century. The splitting of water into hydrogen and oxygen by visible light offers a promising method for the photochemical conversion and storage of solar energy because:

- Hydrogen is an environmentally-valuable fuel, the combustion of which only produces water vapour.
- Hydrogen is easily stored and transported.
- The raw material, water, is abundant and cheap.

Due to its small relative molecular mass, the energy-storage capacity of H_2 is approximately $120 \, kJ \, g^{-1}$; a figure that is almost three times that of the energy-storage capacity of oil.

A semiconductor can act as a photosensitiser, which is characterised by its ability to absorb photons, and then, by utilising the photogenerated electrons and holes, cause the simultaneous oxidation and reduction of reactants.

The photogenerated electrons and holes may recombine in the bulk of the semiconductor or on its surface within a very short time, releasing energy in the form of heat. However, electrons and holes that migrate to the surface of the semiconductor without recombination can respectively reduce and oxidise the reactants adsorbed by the semiconductor. Both surface-adsorption and photochemical-reaction rates are enhanced by use of nano-sized semiconductor particles, as a greatly enhanced surface area is made available.

At the surface of the semiconductor, photogenerated electrons can reduce an electron acceptor, A:

$$A + e^- = A^{\cdot-} \text{ (reduction)}$$

From thermodynamic considerations, in order to reduce a species the potential of the CB of the semiconductor must be more negative than the reduction potential of the acceptor species.

Similarly, photogenerated holes can oxidise an electron donor, D:

$$D + h^+ = D^{\cdot+} \text{ (oxidation)}$$

In order to oxidise a species, the potential of the VB of the semiconductor must be more positive than the oxidation potential of the donor species.

The situation is shown schematically in Figure 11.7, which shows that oxidation and reduction processes can be brought about when the potential values for the CB and VB straddle the potentials of the reduction and oxidation processes.

Fujishima and Honda reported the splitting of water by the use of a semiconductor electrode of titanium dioxide (rutile) connected through an electrical load to a platinum black counter-electrode. Irradiation of the TiO_2 electrode with near-UV light caused electrons to flow from it to the platinum counter-electrode via the external circuit.

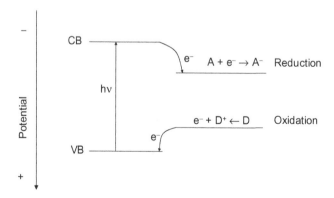

Figure 11.7 The necessary conditions for oxidation and reduction of chemical species brought about by a photoexcited semiconductor. Notice that the electron transfer must always occur in the direction of a more positive (less negative) potential

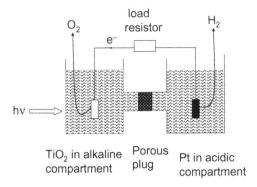

Figure 11.8   Schematic diagrams of the Fujishima–Honda cell using an illuminated TiO_2 (rutile) semiconductor electrode and a platinum counter-electrode

The potential value of the CB for rutile and the H^+/H_2 reduction potential are very similar and an additional driving force was needed to bring about the reduction. This driving force was created by the shifting of the CB electrons of TiO_2 to a more negative potential that was sufficient to produce hydrogen at the platinum electrode in a solution of lower pH.

The TiO_2 electrode was placed in an alkaline solution and the Pt counter-electrode was immersed in an acidic solution, with the two solutions being separated by a porous plug to prevent mixing (Figure 11.8).

Thus, the ability of the holes and electrons to bring about redox chemistry can be controlled by changes in pH which ensure that the CB and VB straddle the hydrogen and oxygen evolution potentials.

The use of the semiconductor strontium titanate(IV), $SrTiO_3$, in a photoelectrochemical cell for the splitting of water was investigated. $SrTiO_3$ has a CB with a more negative potential than rutile and so is able to split water directly, without the need for a pH gradient. Irradiation of the $SrTiO_3$ with near-UV light (388 nm) in the $SrTiO_3/H_2O/Pt$ cell resulted in production of hydrogen and oxygen according to the scheme:

$$\text{Excitation of } SrTiO_2 \text{ by light: } SrTiO_3 + h\nu \rightarrow e^- + h^+$$

$$\text{At the } TiO_2 \text{ electrode: } \frac{1}{2}H_2O \rightarrow \frac{1}{4}O_2 + H^+ + e^-$$

$$\text{At the Pt electrode: } H^+ + e^- \rightarrow \frac{1}{2}H_2$$

$$\text{Overall reaction: } \frac{1}{2}H_2O \rightarrow \frac{1}{2}H_2 + \frac{1}{4}O_2$$

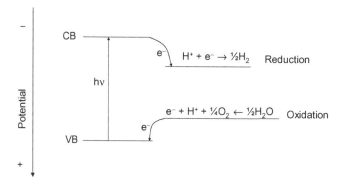

Figure 11.9 Principle of the sensitised water-splitting reaction using SrTiO$_3$ irradiated at 388 nm

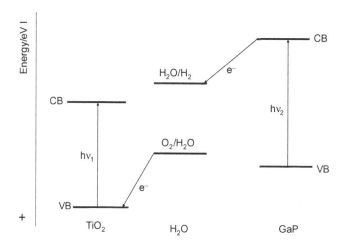

Figure 11.10 Energy-level scheme for the TiO$_2$/GaP cell

In SrTiO$_3$, the potentials of the CB and VB straddle the hydrogen and oxygen evolution potentials (Figure 11.9).

From thermodynamic considerations, all semiconductors that satisfy the above-mentioned energy-gap requirements should be able to be used for photosensitised water splitting. These include TiO$_2$ (anatase form), CaTiO$_3$, Ta$_2$O$_5$ and ZrO$_2$.

An alternative approach involves using two semiconductor electrodes, with the oxidation reaction occurring at an n-type semiconductor (TiO$_2$) and the reduction reaction occurring at a p-type semiconductor (GaP) (Figure 11.10).

This system bears similarity to the so-called Z scheme for photosynthesis (see Chapter 12).

Irradiation of TiO_2 results in an electron being promoted to the CB and an electron from a water molecule being withdrawn into the low-lying vacancy in the TiO_2 VB:

$$\frac{1}{2}H_2O \rightarrow \frac{1}{4}O_2 + H^+ + e^-$$

Similarly, irradiation of GaP promotes an electron to the CB which is then passed to a water molecule:

$$H^+ + e^- \rightarrow \frac{1}{2}H_2$$

11.4   SEMICONDUCTOR PHOTOCATALYSIS

If TiO_2 powder is placed in a small beaker containing polluted water and irradiated with sunlight, a photocatalytic process occurs whereby the pollutants are broken down. The photocatalyst (sensitiser) simultaneously adsorbs the organic pollutant and oxygen, and oxidation and reduction occur on the surface of the semiconductor through the photogenerated holes and electrons, respectively. TiO_2 is an excellent sensitiser for treatment of water on account of its being nontoxic, a good catalyst and inert under the ambient conditions of the process.

Many of the commercial applications of semiconductor photocatalysis involve the oxidative breakdown of organic pollutants in aqueous solution or of volatile organic compounds in air by oxygen, a process called **photomineralisation.**

$$\text{organic pollutant} + O_2 \xrightarrow[hv]{TiO_2} CO_2 + H_2O + O + \text{mineral acids}$$

The processes associated with the photomineralisation of a wide range of organic compounds in aqueous solution using TiO_2 are shown in Figure 11.11, where:

- Photogenerated holes mineralise the organic pollutant. This initially involves the oxidation of surface hydroxyl groups (TiOH) to surface hydroxyl radicals (TiOH$^{\bullet+}$), which then oxidise the organic pollutant and any reaction intermediates.
- Photogenerated electrons reduce oxygen to water, initially generating superoxide (O_2^-), which is further reduced to intermediate species in the formation of water. Hydroxide radicals ($^\bullet$OH) formed as intermediates are also involved in the mineralisation of the organic pollutant.

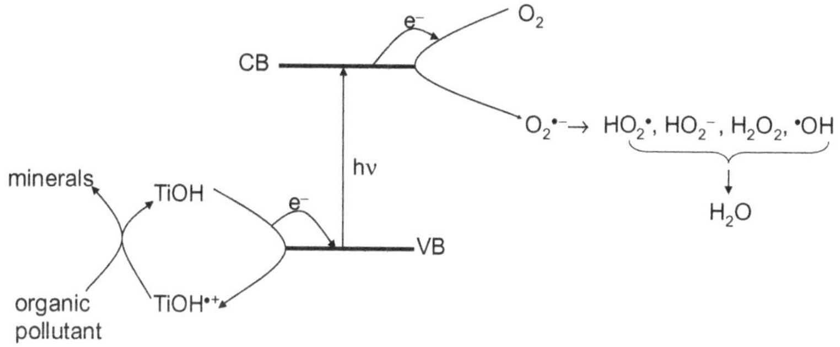

Figure 11.11 Photomineralisation of organic pollutants by oxygen in a process sensitised by a TiO_2 semiconductor
Adapted from A. Mills and S. Le Hunte, 'An Overview of Semiconductor Photocatalysis', *Journal of Photochemistry and Photobiology A*, **108** (1997) © Elsevier

Mineral acids are only formed if heteroatoms such as nitrogen, sulphur and chlorine are present in the pollutant. For example, fenitrothion, an insecticide, undergoes photocatalytic degradation:

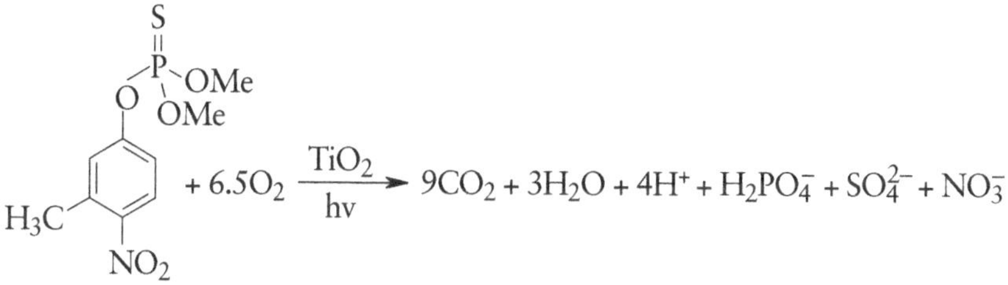

The female sex hormone oestrogen is becoming increasingly prevalent in water supplies due to its use in contraceptive pills and hormone replacement therapy (HRT). This pollutant is associated with many human health problems, such as testicular and breast cancers, infertility and birth defects. Conventional water-treatment processes are not effective against oestrogen pollution but UV irradiation in the presence of titanium dioxide is totally effective in converting the hormone to harmless carbon dioxide and water. This form of treatment is also effective against pesticides, herbicides, detergents and chemicals that mimic oestrogen (xenoestrogens).

Titanium dioxide has also been involved in the photocatalysis of toxic inorganic substances to yield harmless or less-toxic species. Sterilisation of drinking water by chlorine yields potentially carcinogenic compounds so that ozone has been used as an alternative sterilising agent. Bromate ions (a suspected carcinogen) are produced by this process and these can be removed by photosensitisation involving TiO_2:

$$BrO_3^- \xrightarrow[hv]{TiO_2} Br^- + 1.5O_2$$

Heavy metals may be removed from waste water by photosensitisation using TiO_2 where the metal is deposited on the surface of the photocatalyst:

$$Hg^{2+} + H_2O \xrightarrow[hv]{TiO_2} Hg + 2H^+ + \frac{1}{2}O_2$$

A further application for TiO_2 as a semiconductor photocatalyst is in its capacity to act as a mediator in the destruction of biological cells such as pathogens (bacteria, viruses and moulds) and cancer cells. The manner in which the TiO_2 acts is very similar to that found in photomineralisation. Figure 11.12 shows that photogenerated holes produce surface hydroxyl radicals which then oxidise and destroy the cell walls of the biological material so that the cells quickly die. Photogenerated electrons react with adsorbed oxygen at the surface of the semiconductor to form superoxide, which acts as a source of intermediate hydroxyl radicals ($^\bullet$OH).

The use of TiO_2 as a photosensitiser for the destruction of cancer cells involves topical application of the TiO_2 as fine particles to the site of the tumour, which is then irradiated with UV light by means of an optical fibre.

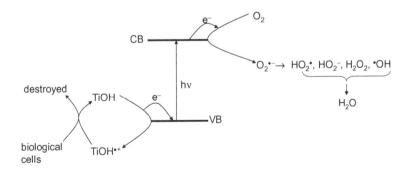

Figure 11.12   Photodestruction of biological cells of pathogens and cancer cells by oxygen in a process sensitised by a TiO_2 semiconductor
Adapted from A. Mills and S-K. Lee, 'A Web-Based Overview of Semiconductor Photochemistry-Based Current Commercial Applications', *Journal of Photochemistry and Photobiology A*, 152(2002) © Elsevier

11.5   SEMICONDUCTOR-PHOTOINDUCED SUPERHYDROPHILICITY

In the absence of any adsorbed species, the photogenerated electron–hole pair in TiO_2 can react with the available surface species; that is, Ti(IV) and bridging O^{2-}.

- Photogenerated electrons reduce Ti(IV) surface species to hydrophilic Ti(III) surface species.
- Oxygen vacancies formed by oxidation of bridging O^{2-} species are generated by photogenerated holes. Water then adsorbs onto the surface, resulting in hydroxylation with an accompanying increase in hydrophilic character.

Thus, the hydrophilic nature of the semiconductor surface is increased on irradiation.

The process is slowly reversed in the dark as the Ti(III) sites are oxidised by atmospheric oxygen and the vacancies are filled by the O^{2-} ions produced by the oxidation reaction (Figure 11.13).

If titanium dioxide is coated on glass then, because of the hydrophilic nature of the irradiated semiconductor, water has a tendency to spread perfectly across the surface. Provided the glass is illuminated, it can undergo self-cleaning by rainfall. Other applications for the hydrophilic glass include antifogging windows and mirrors, where, instead of forming tiny individual water droplets, the surface of the glass is covered by a uniform thin layer of water.

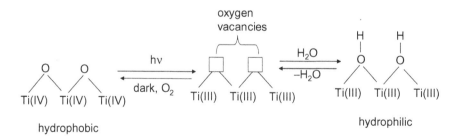

Figure 11.13   Schematic mechanism of the photoinduced superhydrophilicity of TiO_2
Adapted from A. Mills and S-K. Lee, 'A Web-Based Overview of Semiconductor Photochemistry-Based Current Commercial Applications', *Journal of Photochemistry and Photobiology A*, 152(2002) © Elsevier

FURTHER READING

A. Fujishima, T.N. Rao, D.A. Tryk (2000) Titanium dioxide photocatalysis, *J. Photochem. Photobiol. C: Photochem. Reviews*, **1**: 1–21.

M. Grätzel (2005) Solar energy conversion by dye-sensitized photovoltaic cells, *Inorg. Chem.*, **44**: 6841–6851.

K. Hara, H. Arakawa (2003) Dye-sensitized solar cells, In: A. Luque, S. Hegedus (Eds) *Handbook of Photovoltaic Science and Engineering*, Wiley, pp. 663–700.

A. Mills, S. Le Hunte (1997) An overview of semiconductor photocatalysis, *J. Photochem. Photobiol. A: Chemistry*, **108**: 1–35.

12

An Introduction to Supramolecular Photochemistry

AIMS AND OBJECTIVES

After you have completed your study of all the components of Chapter 12, you should be able to:

- Describe the general features of supramolecular photochemistry and how these differ from the features of molecular photochemistry.
- Explain the basis of host–guest photochemistry and be able to give examples of applications involving micelles, zeolites and cyclodextrins.
- Describe the essential photochemical features of the visual process.
- Outline the essential features of the photosynthetic process in green plants and in photosynthetic bacteria.
- Explain how efforts have been made to develop artificial photosynthetic systems. Understand the difficulties and potential benefits of such systems.
- Outline examples of how photochemical supramolecular devices can be used for energy- and electron-transfer processes, for information processing and as light-driven machines.

Principles and Applications of Photochemistry   Brian Wardle
© 2009 John Wiley & Sons, Ltd

12.1   SOME BASIC IDEAS

Probably the highest levels of sophistication found in supramolecular photochemistry are those in the systems involved in photosynthesis and vision. The interaction of light and matter in the photosynthetic and visual processes is characterised by a high degree of structural organisation of the individual molecular components. This is brought about by the various components being assembled in an organised and structurally-integrated manner, resulting in a supramolecular structure.

In its simplest terms, let us consider a model supramolecular system as being a dyad (composed of two components or subunits) A ~ B. From the point of view of a basic definition of supramolecular photochemistry, we may regard this system as being supramolecular if photon absorption by the system results in an electronically-excited state where the excitation is **localised** on a specific component. Likewise, if light absorption leads to electron transfer between the components such that the positive and negative charge are localised on specific components then the system is considered to be supramolecular.

This situation is shown in Scheme 12.1.

The chemical linkage, ~, between A and B shown above can be of any sort, provided modification of the physical and chemical processes of excited states occurs when photoactive components form part of a supramolecular structure.

Thus, several processes may take place within supramolecular systems modulated by the arrangement of the components. These processes involve a modification of the physical and chemical processes of excited states when photoactive components are included in a supramolecular structure. Light excitation of a specific photoactive molecular subunit may also modify the electronic interactions between this subunit and other components, resulting in such phenomena as:

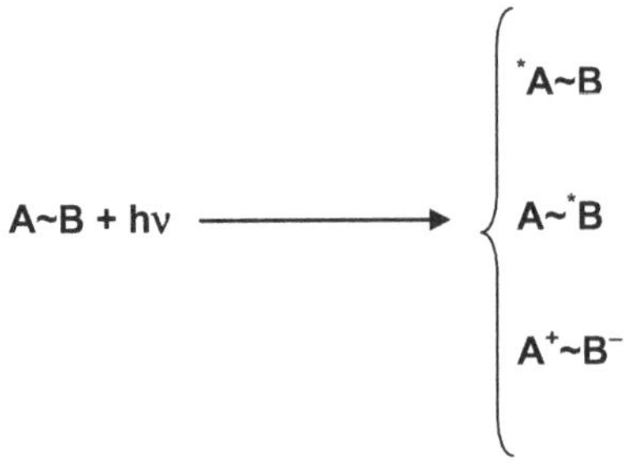

$$A\sim B + h\nu \longrightarrow \begin{cases} {}^*A\sim B \\ \\ A\sim {}^*B \\ \\ A^+\sim B^- \end{cases}$$

Scheme 12.1

- energy transfer or electron transfer between the components

and:

- modification of physical and chemical processes of the excited states of the photoactive components.

Over the last 20 years, supramolecular photochemistry has made considerable advances and photoinduced processes involving supramolecular arrays have been developed that may be exploited for practical purposes, such as:

- light-harvesting antennae
- transfer of the excitation energy to a specific component
- the creation of long-lived charge-separated states
- molecular rearrangement
- reversible on-off switching of luminescence, including molecular sensors
- light-powered molecular machines.

12.2 HOST–GUEST SUPRAMOLECULAR PHOTOCHEMISTRY

Host–guest chemistry describes structures composed of two or more molecules held together in a specific structural relationship by van der Waals forces, hydrogen bonding or ion pairing. The guest molecule must be smaller than the host so as to enable the guest to fit inside the internal cavities of the host. From the point of view of light-induced processes of the guest molecule, the shielding effect of the enveloping host is of major importance. Modification of physical and chemical processes of excited states occurs when photoactive molecules are placed within hosts, such as micelles, cyclodextrins and zeolites.

12.2.1 Micelles

Surfactants are the primary molecular constituents of micelles, each surfactant molecule having separate solvent-loving and solvent-hating groups. Surfactant molecules, such as detergents, self-assemble (aggregate) into roughly spherical-shaped micelles, as shown in Figure 12.1,

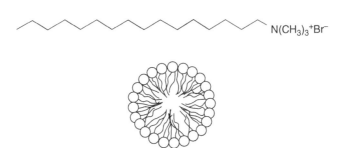

Figure 12.1 Schematic representation of a micelle (bottom) by aggregation of cetyltrimethylammonium bromide ($C_{16}H_{33}N(CH_3)_3^+Br^-$) surfactant molecules (top)

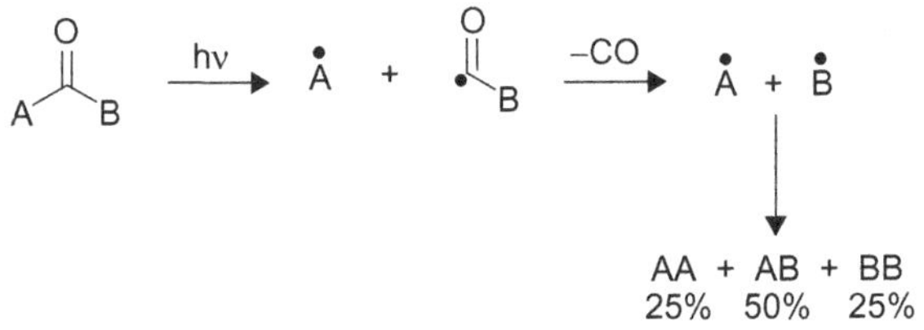

Scheme 12.2

in order to lower the solution free energy. As the surfactant molecules become more concentrated, the interaction between them becomes stronger, until, at a given concentration known as the critical micelle concentration (CMC), supramolecular ordering of the micelle occurs.

Micelles have internal cavities of the order of 1–3 nm diameter, which allow them to act as nanoscale photochemical reactors for incarcerated guest molecules. Photons absorbed by the guest provide the necessary activation to break covalent bonds in the guest molecule, while the resulting reaction intermediates are themselves constrained to remain in the micelle cavity.

Reactions occurring in micelles can give alternative product distributions modified from those encountered in normal reaction pathways:

- The Norrish type 1 photolysis of an asymmetrical ketone A.CO.B normally gives a product ratio of AA:AB:BB as 25:50:25 % (Scheme 12.2).
- In micelles of the surfactant cetyltrimethylammonium chloride (CTAC), the ratio of AA:AB:BB is <1 : >98 : <1 %. The CTAC micelles provide a cage effect, which greatly enhances the joining of the A and B radicals produced by the photolysis (Scheme 12.3).

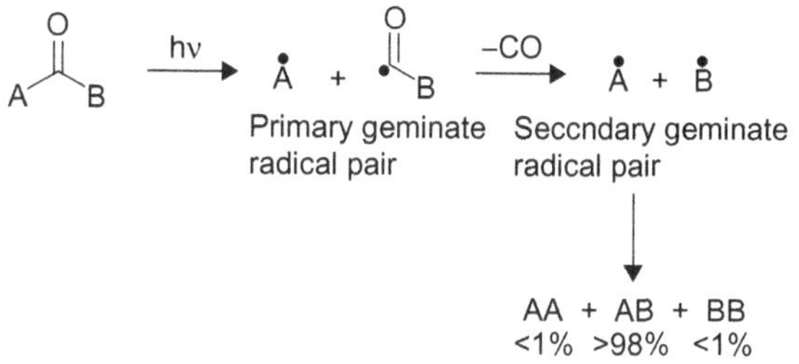

$$A \overset{O}{\underset{}{\|}} B \xrightarrow{h\nu} \overset{\bullet}{A} + \overset{\bullet}{\underset{}{\|}} B \xrightarrow{-CO} \overset{\bullet}{A} + \overset{\bullet}{B}$$

Primary geminate   Secondary geminate
radical pair   radical pair

AA + AB + BB
<1% >98% <1%

Scheme 12.3

The term 'geminate radical pair' is derived from the word 'gemini' = twins, the two geminate radicals being formed together at the same time.

Selective combination of the secondary geminate radical pairs occurs in the micelle, compared to nonselective free-radical combination reactions in solution. This results from the micelle host effectively constraining the separation of the geminate radical pair.

12.2.2   Zeolites as Supramolecular Hosts for Photochemical Transformations

Zeolites are alumino-silicate materials containing extensive channels which connect cavities throughout giant three-dimensional crystalline structures.

Adsorption of compounds may occur on both the external and the internal surfaces of the zeolite, although internal adsorption of a guest molecule is only possible if its size and shape allow it to travel through the internal cavities. So-called 'ship-in-a-bottle' synthesis of guests in zeolite hosts is brought about by using precursor molecules that are small enough to pass into the interior of the zeolite and then react, giving a product that is permanently trapped within the cavities.

Supramolecular concepts involved in the size- and shape-selective aspects of the channels and cavities of zeolites are used to control the selectivity of reactions of species produced by photoexcitation of molecules encapsulated within zeolites. The photochemistry of ketones in zeolites has been extensively studied. Photoexcitation of ketones adsorbed on zeolites at room temperature produces radical species by the Norrish type 1 reaction. A 'geminate' (born together) radical pair is initially produced by photolysis of the ketone, and the control of the reaction products of such radicals is determined by the initial supramolecular structure

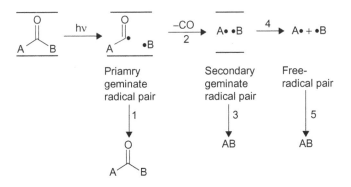

Scheme 12.4

of the ketone–zeolite complex and by the possible pathways taken as the radicals move through the zeolite framework.

In such photochemical reactions, bonds are broken by the absorption of light, which causes the formation of reactive intermediates (radicals) without any activation of the zeolite (Scheme 12.4).

Photochemical excitation results in α-cleavage to produce a **primary, geminate radical pair**, which may undergo radical combination reactions (1) in competition with decarbonylation or (2) to produce a **secondary geminate radical pair**. The latter may undergo radical combination (3) or produce a free-radical pair (4). The free radicals undergo radical combination reactions (5).

In solution, the Norrish type 1 reaction of ketones results in the nonselective free-radical combination reactions to give products AA, AB and BB in the ratio of 1:2:1, whereas photolysis of ketones in zeolites produces:

- Selective stereochemical and regiochemical combination of the geminate primary pairs.
- Selective combination reactions of geminate secondary pairs.
- Selective combination of free-radical pairs.

Regioselectivity is the preferential formation of one product over all other possibilities. The zeolite host effectively constrains separation of the secondary geminate radical pair A• and B•.

Thus regioselective combination of this radical pair occurs in the zeolite, compared to nonselective free-radical combination reactions in solution.

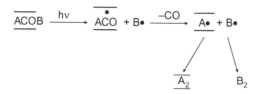

Scheme 12.5

Table 12.1 Characteristic wavelengths of ultraviolet light

Name	Wavelength (nm)
UV-A	400–320 ·
UV-B	320–280
UV-C (absorbed by the ozone layer in the stratosphere)	280–100

If the molecular geometry of the ketone, ACOB, does not allow its complete inclusion within the zeolite because the bulkier part of the ketone (say B) is unable to fit within the cavities then photolysis followed by decarbonylation will result in two distinct products, AA and BB (Scheme 12.5).

- The internal A• radicals will diffuse through the channel system and combine with other A• radicals.
- The size-excluded B• radicals on the surface of the zeolite are able to combine exclusively with other B• radicals.

When radicals with bulky substituents are found within the internal channels of the zeolite, their interactions may be inhibited by the narrow diameter of the channels, allowing them to persist for a long time (shown by electron spin resonance (ESR) spectroscopy). These steric constraints are known as a **supramolecular steric effect**.

Zeolites have recently found application in the field of sunscreens, which are organic compounds acting as efficient filters of UV-A and UV-B (Table 12.1).

As public consciousness of the harmful effects of exposure to ultraviolet light (skin cancer, premature ageing, etc.) has increased, there has been a substantial increase in the use of sunscreens. These sunscreens have, however, been associated with a number of reported problems such as phototoxicity and photoallergy. The use of p-aminobenzoic acid (PABA) and its derivatives has been discontinued due to their ability to sensitise the formation of singlet oxygen and cause damage to DNA. Encapsulation of PABA within zeolites has been shown to have no effect on its beneficial photoprotecting properties while resulting in a significant decrease in its harmful effects.

Insoluble sunscreens, where an organic filter molecule is mixed with a light-scattering metal oxide such as TiO_2 or ZnO, have become increasingly popular. However, it would be expected that, because of the photocatalytic properties of the metal oxides, the organic material would undergo photoinduced degradation and lose its effectiveness as a UV filter. Encapsulation of the organic filter in zeolites has been shown to preserve the stability of the organic sunscreen and protect it from photocatalytic degradation. Such a supramolecular sunscreen retains all the benefits of the organic filter but prevents its interaction with both the skin and the metal oxide.

12.2.3 Cyclodextrins as Supramolecular Hosts

Cyclodextrins (Table 12.2) are cyclic molecules made up of glucose monomers coupled to form a rigid, hollow, tapering torus with a hydrophobic interior cavity (Figure 12.2). Because of the presence of the cavity, cyclodextrins are able to act as hosts, binding with small guest molecules held within the internal cavity.

Of particular interest in the application of cyclodextrins is the enhancement of luminescence from molecules when they are present in a cyclodextrin cavity. Polynuclear aromatic hydrocarbons show virtually no phosphorescence in solution. If, however, these compounds in solution are encapsulated with 1,2-dibromoethane (enhances intersystem crossing by increasing spin–orbit coupling ~ external heavy atom effect) in the cavities of β-cyclodextrin and nitrogen gas passed, intense phosphorescence emission occurs at room temperature. Cyclodextrins form complexes with guest molecules, which fit into the cavity so that the microenvironment around the guest molecule is different from that in

Table 12.2 Composition of cyclodextrins

Cyclodextrin	Number of Glucose Units in the Ring
α-cyclodextrin	6
β-cyclodextrin	7
γ-cyclodextrin	8

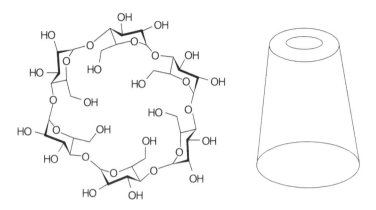

Figure 12.2 Schematic representations of an α-cyclodextrin molecule showing the glucose rings and the overall topology

the bulk medium. Once the phosphorescent guest molecule occupies the cavity, other molecules are excluded from simultaneously occupying it and collisional deactivation of the excited triplet state of the guest molecule is reduced or prevented.

12.3 SUPRAMOLECULAR PHOTOCHEMISTRY IN NATURAL SYSTEMS

The two most important photochemical processes that take place in the natural world are photosynthesis and vision, in which photons are utilised for entirely different roles. Light is used by living organisms as energy in photosynthetic processes and as information in visual processes. Evolutionary processes have brought about a high level of sophistication where highly-organised arrangements of molecular components are assembled into photochemical supramolecular devices capable of carrying out complex biological functions.

12.3.1 Vision

The rod and cone cells found in the retina of the eye are functional supramolecular devices involved in information processing. Rod cells function in dim light and are black and white receptors while cones are colour receptors.

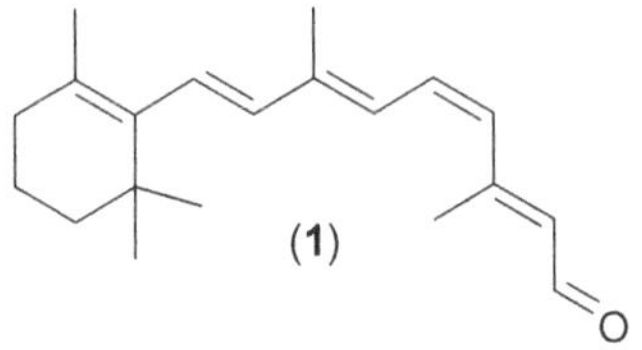

Figure 12.3   11-cis-retinal (1) and all-trans-retinal (2)

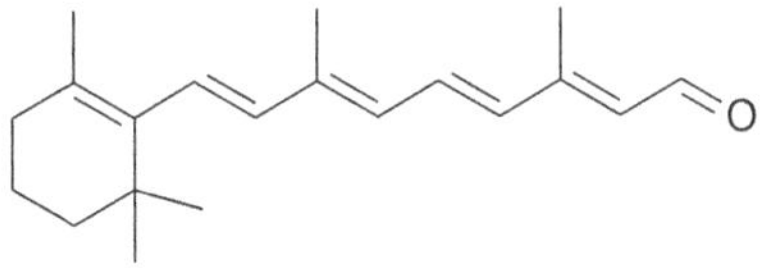

The photochemistry of vision provides us with an example of **host–guest supramolecular photochemistry** where the smaller 11-cis-retinal guest molecule is held within the internal cavity of the much larger protein host molecule (opsin) as a result of noncovalent bonding.

The light-absorbing part of a rod cell contains the pigment **rhodopsin**, which consists of the opsin attached to the 11-cis-retinal molecule (**1**) (Figure 12.3). Free 11-cis-retinal absorbs in the ultraviolet, but when attached to opsin the absorption is in the visible region.

After absorbing a photon, the 11-cis-retinal undergoes photoisomerisation into its geometric isomer all-trans-retinal (**2**) (Figure 12.3).

Because of this photoisomerisation, structural changes occur within the confines of the binding cavity, which in turn produce changes in the opsin and the attached cell-membrane protein. This results in functional changes to the cell membrane, culminating in generation of a signal impulse which is sent to the brain.

In colour vision there are three specific types of cone cell corresponding to red, green and blue receptors. The chromophore is the same for all three colours, being 11-cis-retinal bound to a protein which is structurally similar to opsin. Colour selectivity is achieved by positioning specific amino acid side chains along the chromophore so as to perturb the absorption spectrum of the chromophore.

12.3.2 Photosynthesis

Photosynthesis is the process whereby green plants and certain micro-organisms capture solar energy and convert it to carbohydrates and other complex molecules. In all cases except for photosynthetic bacteria, the overall reaction is written as:

$$nCO_2 + nH_2O + hv \rightarrow (CH_2O)_n + nO_2; \Delta G = +500 kJ mol^{-1}(CO_2)$$

Thermodynamically, since the free-energy change has a positive value it can only proceed when energy is supplied from the sun. The photosynthetic process occurs on a vast scale, fixing 2×10^{11} tonnes per annum of carbon as carbohydrate, representing 10 times the energy consumption of the entire human race.

The photosynthetic process thus provides us with an example of a complex, light-powered **photochemical molecular device** which uses light as an energy supply in order to facilitate energy conversion. In green plants, this molecular device is located within a specially-adapted **photosynthetic membrane**.

Photosynthesis occurs in two stages:

- **Light reactions**: light energy is converted into short-term chemical energy, producing oxygen as a by-product.
- **Dark reactions**: the short-term chemical energy is used to convert carbon dioxide into carbohydrate. We will not be concerned with these dark reactions.

The leaves of plants are green because they contain the primary light-absorbing pigments called the **chlorophylls**, which absorb strongly in the blue and red regions of the visible spectrum, leaving the intermediate green wavelengths to be reflected to our eyes (Figure 12.4).

The chlorophyll molecule (Figure 12.5) consists of:

- A rigid, planar, conjugated **porphyrin ring**, which functions as an efficient absorber of light.
- A hydrophobic **phytol chain**, which keeps the chlorophyll molecule embedded in the photosynthetic membrane.

Accessory pigments called **carotenoids**, such as β-carotene (3) (Figure 12.6), are also found in plants. These contain an extended conjugated system and absorb principally in the blue and green regions of the visible spectrum, while reflecting the red, orange and yellow wavelengths.

Carotenoids are needed for the survival and productivity of the plant:

- Carotenoids complement chlorophylls in the light-harvesting process.

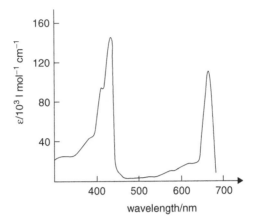

Figure 12.4 The absorption spectrum of a solution of chlorophyll a in ether

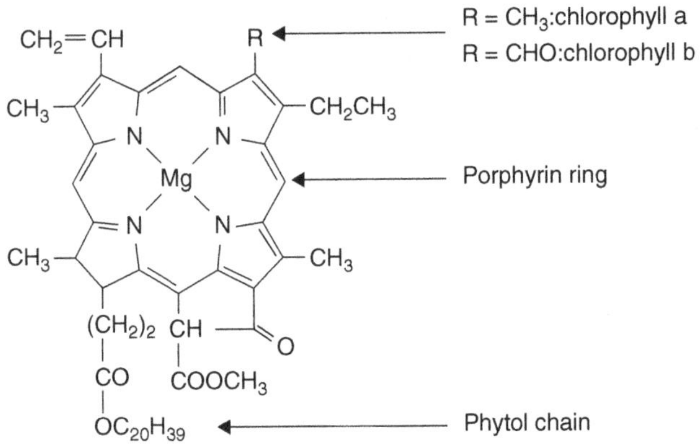

CH$_2$=CH   R   R = CH$_3$:chlorophyll a
     R = CHO:chlorophyll b

CH$_3$ CH$_2$CH$_3$

N   N

Mg   Porphyrin ring

N   N

CH$_3$ CH$_3$

(CH$_2$)$_2$ CH

CO   COOCH$_3$

OC$_{20}$H$_{39}$ Phytol chain

Figure 12.5 Structure of chlorophyll a and chlorophyll b molecules

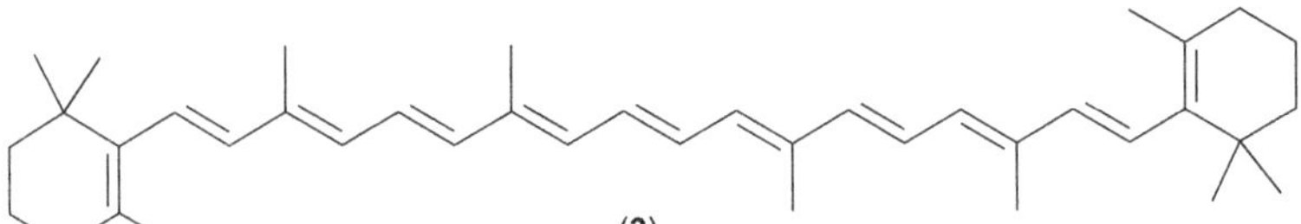

(3)

Figure 12.6 β-carotene

- The structure and assembly of the photosystems are dependent on the availability of specific carotenoids that assist in the correct folding and maintain stability of the photosystem proteins.
- Carotenoids are essential for the photoprotective mechanism employed by plants to dissipate excess photon energy absorbed by chlorophyll as heat, thus preventing formation of highly-reactive oxygen species. Carotenoids also deactivate singlet oxygen ($^1O_2$) generated in the antennae and reaction centres.

The sequence of light reactions responsible for photosynthesis may be considered as:

1. **Light harvesting**: several hundred pigment molecules act together as the **photosynthetic unit** (Figure 12.7), which is made up from the **light-harvesting antenna** and the **reaction centre**, consisting of a chlorophyll dimer.

   The complex arrangement of pigment molecules held together in protein complexes by intermolecular forces acts as the light-harvesting antenna that allows the absorption of a broad range of wavelengths. The antenna also provides the means whereby rapid stepwise transfer of the excitation energy to the reaction centre is achieved. Energy is repeatedly passed between adjacent pigment molecules such that each successive step involves energy transfer

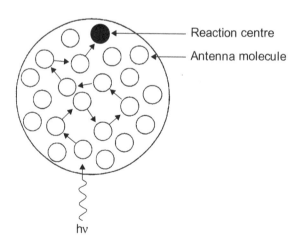

Figure 12.7 The photosynthetic unit, in which an antenna chlorophyll molecule is excited by photon absorption and the energy is transferred to the chlorophyll dimer at the reaction centre

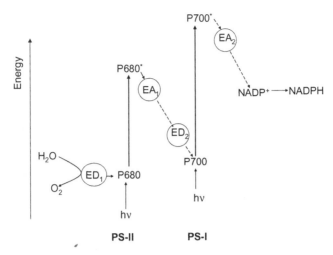

Figure 12.8 The Z scheme, an overview of the flow of electrons during the light-dependent reactions of photosynthesis. ED and EA refer to the electron donors and acceptors of the two photosystems, respectively

to a pigment molecule that absorbs light of equal or longer wavelength (lower energy).

The energy transfer occurs by means of the Coulombic long-range mechanism (Section 6.6), which ultimately redistributes the excitation energy via the adjacent pigment molecules to the reaction centre. Excitation of the reaction centre is over within a few femtoseconds.

2. **Charge separation and transport**: the electrons are transferred through an electron transport chain, the so-called **Z scheme** (Figure 12.8). The Z scheme is made up of two principal parts, **photosystem I** (PS-I) and **photosystem II** (PS-II). The reaction centre of PS-II is a chlorophyll dimer referred to as **P680**, P meaning pigment and 680 being the wavelength in nanometres at which this dimer absorbs most strongly. The reaction centre of PS-I is also a chlorophyll dimer, designated **P700**.

The first chemical steps occur a few picoseconds after absorption, when:

- Excited P680* loses an electron to the electron acceptor (EA_1) in PS-II, producing $P680^+$ and EA_1^-:

$$P680 \xrightarrow{\ h\nu\ } P680^* \underset{EA_1 \quad EA_1^-}{\xrightarrow{\hspace{2cm}}} P680^+$$

- Excited P700* loses an electron to the electron acceptor (EA_2) in PS-II, producing P680⁺ and EA_2^--excited P700*:

$$P680 \xrightarrow{\text{hv}} P680^* \xrightarrow{\hspace{2cm}} P680^+$$
$$EA_1 \quad EA_2^-$$

These are the only steps at which light energy is converted to chemical energy. The remaining electron transport steps involve a decrease in the free energy of the system, where the free energy released can be utilised as useful chemical energy in the production of NADPH, which is necessary for producing carbohydrates from CO_2 in the dark reactions.

The supramolecular structure of the photosystems ensures the membrane location of the charged species is such that charge recombination between EA_1^- and P680⁺ and between EA_2^- and P700⁺ is prevented.

3. **Water oxidation:** we saw above that the absorption of light by PS-II leads to the formation of two charged molecules, P680⁺ and EA_1^-. P680⁺ is so strongly oxidising that it is capable of being rapidly reduced by electrons drawn from water. This reaction occurs at a manganese-containing protein which acts as a donor of electrons (ED_1) to P680⁺.

$$\frac{1}{2}H_2O + hv \rightarrow H^+ + \frac{1}{4}O_2 + e^-$$

The water acts as a large reserve of electrons for the photosynthetic electron transport process in the Z scheme, with the oxygen being produced as a waste product.

12.3.3   Bacterial Photosynthesis

A simpler and better-understood process of the primary photosynthetic reaction and charge separation occurs in bacterial photosynthesis, which has only one photosystem instead of the two photosystems of green-plant photosynthesis.

X-ray studies of the reaction centre have led to a detailed understanding of its structure (Figure 12.9).

The following processes occur in bacterial photosynthesis:

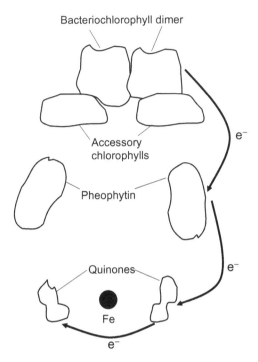

Figure 12.9   Schematic view of the bacterial-photosynthetic reaction centre and the energy transfers which occur. The groups are held in a fixed geometry by the surrounding proteins

- Sunlight is efficiently absorbed throughout the solar spectrum by an antenna system containing bacteriochlorophyll and carotenoid chromophores in a protein environment.
- Rapid multistep Coulombic energy transfer takes place as the excitation energy is transferred between the antenna chromophores and the 'special pair' of bacteriochlorophyll molecules (P) in the reaction centre.
- There is rapid (~ps) electron transfer to a bacteriopheophytin molecule (BP).
- There is charge separation by electron transfer from bacteriopheophytin to a quinine (Q_A) and then on to a second quinine (Q_B):

$$P.BP.Q_A.Q_B \xrightarrow{h\nu} P^*.BP^-.Q_A.Q_B \rightarrow P^+.BP^-.Q_A.Q_B \rightarrow P^+.BP.Q_A^-.Q_B \rightarrow P^+.BP.Q_A.Q_B^-$$

- After reduction of the oxidised 'special pair' by a c-type cytochrome, the energy of a second photon is used to transfer a second electron to Q_B:

$$Q + 2e^- + 2H^+ = QH_2$$

- The resulting hydroquinone (QH_2) then diffuses to the cytochrome bc complex, which oxidises QH_2 back to Q, using the resulting reduction potential, via cytochrome c, to reduce the 'special pair' and hence regenerate the reaction centre.
- The energy released is used to transfer protons across the photosynthetic membrane and ultimately this energy acts as a driving force for the catalysed production of high-energy adenosine triphosphate (ATP) from adenosine diphosphate (ADP) and inorganic phosphate.

12.4   ARTIFICIAL PHOTOSYNTHESIS

We have seen that, in photosynthetic bacteria, visible light is harvested by the antenna complexes, from which the collected energy is funnelled into the 'special pair' in the reaction centre. A series of electron-transfer steps occurs, producing a charge-separated state across the photosynthetic membrane with a quantum efficiency approaching 100%. The nano-sized structure of this solar energy-conversion system has led researchers over the past two decades to try to imitate the effects that occur in nature.

Synthetic chemistry enables us to mimic the energy- and electron-transfer processes by linking together donor and acceptor groups by means of covalent bonds or bridging groups, rather than using the protein matrix found in natural systems.

Because multistep electron transfer to achieve long-lived charge separation is necessary in natural photosynthesis, it might be considered that the intelligent approach toward designing artificial systems would be to bring about similar processes. The simplest system capable of carrying out such operations is one composed of a covalently-linked **triad** (a three-component system). Various triads consisting of a chromophore (C), an electron acceptor (A) and an electron donor (D) have been investigated (Figure 12.10).

Excitation of the chromophore is followed by photoinduced electron transfer to the electron acceptor, then electron transfer from the electron donor to the oxidised chromophore:

$$\text{D.C.A} \xrightarrow{h\nu} \text{D.C}^*\text{.A} \xrightarrow{e^-} \text{D.C}^+\text{.A}^- \xrightarrow{e^-} \text{D}^+\text{.C.A}^-$$

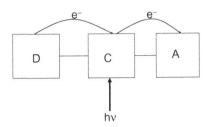

Figure 12.10   Triad capable of producing photoinduced charge separation

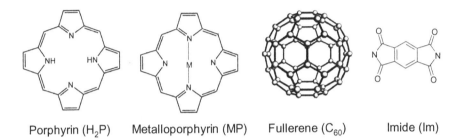

Porphyrin (H₂P)   Metalloporphyrin (MP)   Fullerene (C₆₀)   Imide (Im)

Figure 12.11   Component building blocks used in triads for solar energy-conversion systems

The lifetime of $D^+.C.A^-$ produced by the sequential electron transfer increases as the distance between D^+ and A^- increases. Porphyrin-imide-fullerene triads have been synthesised (Figure 12.11), with the three components separated by nonreactive spacer molecules.

In artificial photosynthetic models, porphyrin building blocks are used as sensitisers and as electron donors while fullerenes are used as electron acceptors. Triads, tetrads, pentads and hexads containing porphyrins and C_{60} have been reported in the literature (see the Further Reading section).

$$ZnP.Im.C_{60} \xrightarrow{h\nu} ZnP^*.Im.C_{60} \longrightarrow ZnP^+.Im^-.C_{60} \longrightarrow ZnP^+.Im.C_{60}^-$$

Furthermore, a pentad consisting of a caretenoid (C), zinc porphyrin (ZnP), porphyrin (H₂P) and two quinines (QA and QB) has been produced, which aims to utilise components allied to those found in natural photosynthetic systems. The zinc porphyrin absorbs at 650 nm, a process that is followed initially by fast energy transfer:

$$\text{C.ZnP.H}_2\text{P.Q}_A\text{.Q}_B \xrightarrow{h\nu} \text{C.ZnP}^*.\text{H}_2\text{P.Q}_A\text{.Q}_B \xrightarrow{\text{Energy transfer}} \text{C.ZnP.H}_2\text{P}^*.\text{Q}_A\text{.Q}_B$$

Electron transfer follows from H_2P^* to Q_A, and then a charge-separation sequence such that localised charges reside on the carotenoid and on the quinine (Q_B):

$$\text{C.ZnP.H}_2\text{P}^*.\text{Q}_A\text{.Q}_B \xrightarrow{e^-} \text{C.ZnP.H}_2\text{P}^+.\text{Q}_A^-.\text{Q}_B$$

Charge
separation

$$\text{C}^+.\text{ZnP.H}_2\text{P.Q}_A\text{.Q}_B^-$$

In the supramolecular systems above, following the charge-separation process, part of the absorbed light energy is stored as short-term redox energy. In natural photosynthesis, this energy is converted into high-energy chemical substances (fuels). From the point of view of an artificial photosynthetic system, the most attractive fuel-producing reaction is the cleavage of water.

From a consideration of thermodynamics, solar energy conversion by means of photoinduced charge separation followed by water splitting is a feasible process.

For the donor–chromophore–acceptor triad of Figure 12.10:

$$\text{D.C.A} \xrightarrow{h\nu} \text{D.C}^*.\text{A} \xrightarrow{e^-} \text{D.C}^+.\text{A}^- \longrightarrow$$

$$\text{D}^+.\text{C.A}^- + \text{H}_2\text{O} \longrightarrow \text{D}^+.\text{C.A} + \frac{1}{2}\text{H}_2 + \text{OH}^-$$

$$\text{D}^+.\text{C.A} + \frac{1}{2}\text{H}_2\text{O} \longrightarrow \text{D.C.A} + \frac{1}{4}\text{O}_2 + \text{H}^+$$

Because the charge separation is a one-electron process but the water-splitting reactions are multi-electron processes (although they have been written above as one-electron processes for simplicity), suitable catalysts are needed to accelerate these multi-electron processes so they can be brought about during the lifetime of the photoinduced species.

Despite much effort to address the thermodynamic and kinetic impli-
cations of the photochemical water-splitting reaction, real systems in
which the reduction and oxidation processes occur together have not
met with a great deal of success. As we might expect, the two-electron
photoreduction of water is easier to achieve than the four-electron pho-
tooxidation. Thus much research has concentrated on the photoreduction process, which has the added advantage of producing hydrogen (a
'clean' fuel that only produces water on combustion). In order that the
catalytic production of hydrogen will proceed, a sacrificial electron
donor is used as an external source of electrons.

Successful systems have used colloidal platinum as an efficient catalyst
for the multi-electron reduction process by which hydrogen is produced.
The platinum acts as a 'charge pool' in that electrons from one-electron
processes are trapped, to be later delivered to the substrate in a concerted manner, thus avoiding formation of high-energy intermediates
(Figure 12.12).

Other successful systems employed for the reduction of water to
hydrogen are given in Table 12.3.

For more details regarding artificial photosynthesis systems, the reader
is referred to Section 7.5.

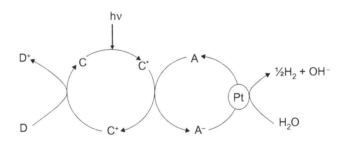

Figure 12.12 Building blocks of an artificial photosynthesis system for hydrogen
production using a chromophore (C), an electron acceptor (A) and a sacrificial
donor (D)

Table 12.3 Compounds that have been used as components in successful systems
for the photoreduction of water

Sacrificial Donors	Acceptors	Chromophores	Catalysts
Triethanolamine (TEOA)	Methyl viologen (MV^{2+})	Chlorophyll	Colloidal Pt
EDTA		$Ru(bpy)_3^{2+}$	Hydrogenase
Cysteine		Metal porphyrins	

12.5 PHOTOCHEMICAL SUPRAMOLECULAR DEVICES

A device is something that performs a useful function, and the idea of a device can be extended to the molecular level by designing and synthesising supramolecular species that are able to perform specific functions. These photochemical supramolecular devices require absorption of light energy to cause electronic or nuclear rearrangements, and often luminescence is used to monitor their operation. In the long term, it is expected that such devices will find applications in the field of information processing and lead to the construction of chemical-based computers.

Photochemical supramolecular devices can be conveniently divided into three groups.

12.5.1 Devices for Photoinduced Energy or Electron Transfer

See Figure 12.13.

Molecular wires conduct an electrical signal (which could be just one electron) between two connected components over a long distance. This function can be brought about by linking a donor and acceptor by means of a rigid spacer.

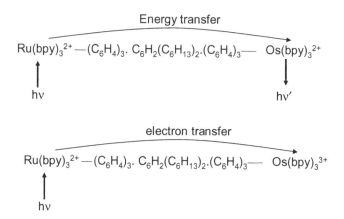

Figure 12.13 Photoinduced energy transfer and electron transfer over long distances
Adapted from V. Balzani, 'Photochemical molecular devices', *Photochemical and Photobiological Sciences*, 2003 (2), 459–476

Dendrimers are highly-branched macromolecules with large tree-like structures, which can be attached to simpler metal complexes. Photoactive dendrimers have been synthesised around a $Ru(bpy)_3^{2+}$ core with peripheral naphthyl units. In acetonitrile solution, this leads to enhanced fluorescence compared to the basic ruthenium complex, showing that very efficient energy transfer is taking place, converting the short-lived UV fluorescence of the naphthyl units to a long-lived orange emission of the metal core (antenna effect). This dendrimer has various applications, such as antenna systems for harvesting energy in sunlight.

12.5.2 Devices for Information Processing based on Photochemical or Photophysical Processes

Logic gates are devices that can perform basic logic operations, depending on the nature of the input and output signals. The relation between the inputs and outputs is summarised in the **truth table** of the logic gate.

For example, the **AND** logic gate has two inputs and one output. The truth table for the AND gate is shown in Table 12.4.

There is an output signal (1) only when there is an input 1 signal (1) AND an input 2 signal (1).

A molecular-level two-input AND gate can be made from two covalently-linked receptors, R_1 and R_2, that are able to quench the luminescence of a fluorophore, F. When one of the receptors acts as host to a suitable species, it can no longer quench the luminescence of F, but the luminescence can be quenched by the other free receptor. The output signal (luminescence from F) can only be observed when both receptors are bound to chemical species (Figure 12.14).

In compound (4) (Figure 12.15) the fluorescence of the anthracene group is quenched by electron transfer from both the crown ether unit and the amino group. The electron transfer processes from the crown ether and the amino group can be prevented by adding Na^+ and H^+, respectively.

Table 12.4 Truth table for the AND logic gate

INPUT 1	INPUT 2	OUTPUT
0	0	0
0	1	0
1	0	0
1	1	1

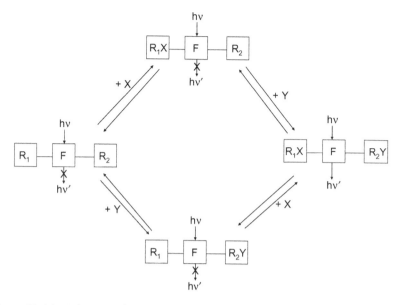

Figure 12.14 Schematic diagram for the operation of a two-input molecular AND gate. R_1 and R_2 are receptors for chemical species X and Y, respectively

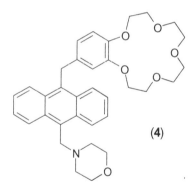

Figure 12.15 Compound (4)

12.5.3 Devices Designed to Undergo Extensive Conformational Changes on Photoexcitation: Photochemically-driven Molecular Machines

Mechanical movements in supramolecular structures rely on modulation of noncovalent-bonding interactions. Such changes occur when charge-transfer interactions between electron-donor and electron-acceptor groups are weakened.

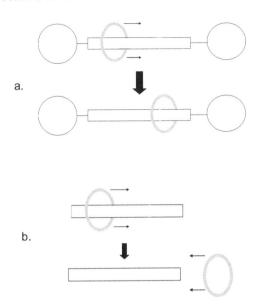

Figure 12.16 Schematic representation of (a) ring shuttling in rotaxanes and (b) dethreading, rethreading in pseudorotaxanes

Rotaxanes are mechanically-linked molecules consisting of a molecular strand with a cyclic molecule linked around it. The molecular strand is terminated with bulky end groups at both ends. Rotaxanes find use in the fact that there are a number of positions (stations) along the molecular strand to which the cyclic molecule can temporarily attach.

Pseudorotaxanes have the general form of a rotaxane but the molecular-strand component is without bulky end groups.

Figure 12.16 shows a schematic representation of intercomponent motions that can be obtained in pseudorotaxanes and rotaxanes.

Photochemically-driven dethreading and rethreading of an azobenzene-based pseudorotaxane in acetonitrile occurs when (**5**) threads with (**6**) (Figure 12.17).

The cis isomer of (**5**) does not form a pseudorotaxane with (**6**), a fact that is exploited to obtain photoinduced (365 nm) dethreading of the pseudorotaxane by trans → cis isomerisation of (**5**). When the cis isomer is converted back to the trans form by light irradiation (436 nm), the pseudorotaxane is obtained once again.

Design of a photochemically-driven linear motor can be brought about by use of a rotaxane in which the ring component can be moved between different stations by means of light absorption (Figure 12.18).

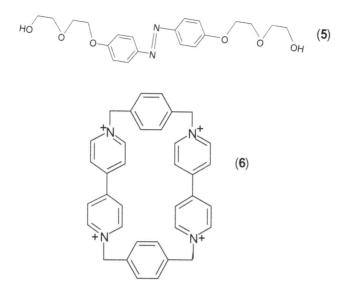

Figure 12.17 Compounds (5) and (6)

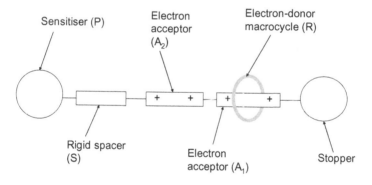

Figure 12.18 Schematic representation of a linear motor powered by light Adapted from V. Balzani, A. Credi and M. Venturi, 'Light-powered molecular-scale machines', *Pure and Applied Chemistry* Volume 75, No. 5, 541–547 © International Union of Pure and Applied Chemistry IUPAC 2003

The components of Figure 12.18 are assembled from the chemical structures shown in Figure 12.19.

The mechanism by which the motion occurs is based on the following four steps:

- **Destabilisation of the A_1 station:** light absorption by P is followed by electron transfer from the excited state to the A_1 station while P becomes oxidised to P^+.

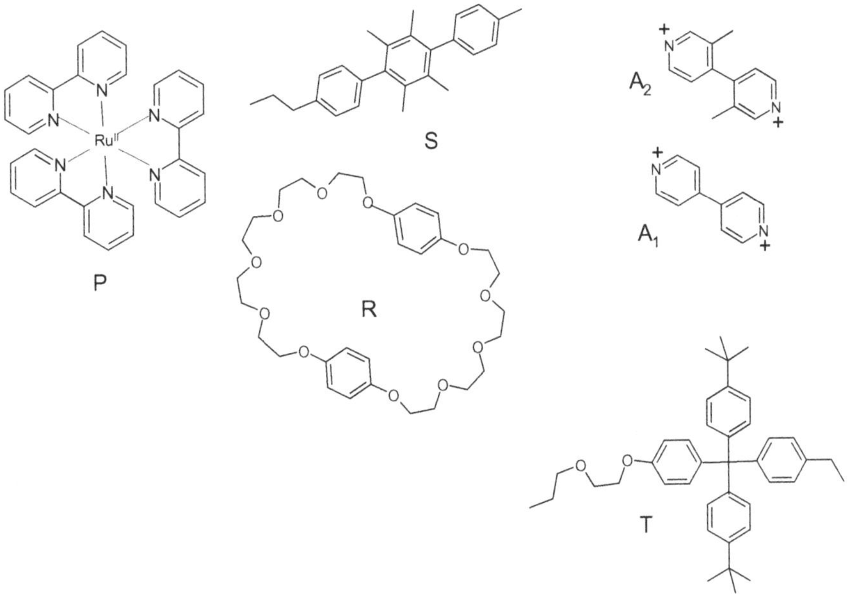

Figure 12.19 Component molecular structures that are assembled to give the device shown in Figure 12.18
Adapted from V. Balzani, A. Credi and M. Venturi, 'Light-powered molecular-scale machines', *Pure and Applied Chemistry* Volume 75, No. 5, 541–547 © International Union of Pure and Applied Chemistry IUPAC 2003

- **Ring displacement:** the ring moves from the reduced A1 station to A$_2$.
- **Electronic reset:** back electron transfer from the reduced station A$_1$ to P$^+$ resets the electron acceptor power to the A$_1$ station.
- **Nuclear reset:** because of the electronic reset, the ring moves from A$_2$ to A$_1$.

FURTHER READING

N. Armaroli (2003) From metal complexes to fullerene arrays: exploring the exciting world of supramolecular photochemistry fifteen years after its birth, *Photochem. Photobiol. Sci.*, **2**: 73–87.

V. Balzani (2003) Photochemical molecular devices, *Photochem. Photobiol. Sci.*, **2**: 495–476.

V. Balzani, A. Credi, F.M. Raymo, J. Fraser Stoddart (2000) Artificial molecular machines, *Angew. Chem. Int. Ed.*, **39**: 3348–3391.

V. Balzani, A. Credi, S. Silvi, M. Venturi (2006) Artificial nanomachines based on inter-locked molecular species: recent advances, *Chem. Soc. Rev.*, **35**: 1135–1149.

V. Balzani, A. Credi, M. Venturi (2003) Light-powered molecular-scale machines, *Pure Appl. Chem.*, **75**: 541–547.

Index

Note: Page numbers in *italics* refer to Figures; those in **bold** to Tables.